GASTERIAS

OF SOUTH AFRICA

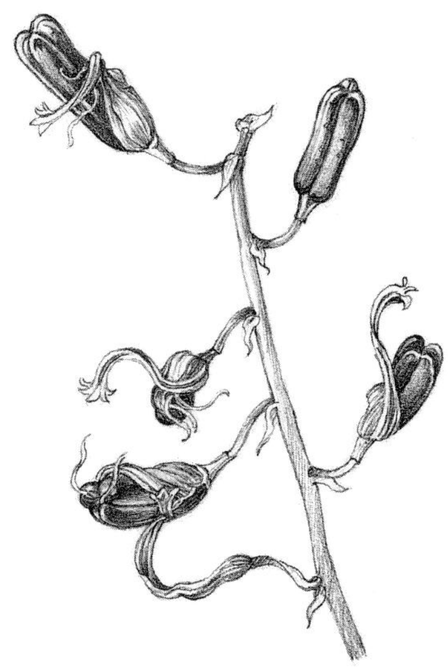

1
2
3
4.
4/1984.

GASTERIAS

OF SOUTH AFRICA

A new revision of a major succulent group

Ernst J. van Jaarsveld

Illustrations

Ellaphie Ward-Hilhorst

FERNWOOD PRESS

in association with the
National Botanical Institute

ACKNOWLEDGEMENTS

First and foremost I acknowledge my indebtedness to our father in heaven, creator of all life. I am also indebted to the late Mr Frank Stayner, to whom this book is dedicated. His personal communications and his gift of *Gasteria* publications, reprints, colour slides and photographs, collected over many years until his death in 1981, were the inspiration behind this study.

I wish to express my gratitude to Professor Brian Huntley and Professor Kobus Eloff for their encouragement in this work and for giving me the opportunity to visit European botanical gardens and herbaria. I am especially grateful to Professor Charles Stirton for his guidance, positive criticism, encouragement and the many constructive discussions we had during the course of my *Gasteria* research.

To Ellaphie Ward-Hilhorst, my sincere thanks for editing parts of this book when it was in manuscript form and also for her enthusiastic encouragement.

Practical and professional assistance and advice were generously given by Dr John Rourke, Dr Leslie Codd and Mr Bruce Bayer, and I thank them most warmly.

My thanks too to Mr Tommy Sardien and Mr Eric Marthinus for their care and cultivation of the large collection of living plants housed at Kirstenbosch. Brenda Hickey is also thanked for assisting in the artificial hybridization of *Gasteria* species.

For their invaluable and unstinted support at all times, I am indebted to the late Professor Ted Schelpe and Mrs D. Rex of the Bolus Herbarium, Mr George McKrill and also my father, Professor Floors van Jaarsveld.

To Professor Onno Wijnands and Mr Ben Groen of the Agricultural University at Wageningen, and to Mr Brian Schrire, South African liaison officer at the Royal Botanic Gardens, Kew, who kindly made available photocopies of the Moninckx Atlas and other reprints, a special word of gratitude is due.

The following people assisted me with live material and I am much indebted to them: G. Nichols of Durban; T. Abbott of Umtamvuna; K. Kritzinger of Vanrhynsdorp; B. Branch of King William's Town; H. Gie, J. Venter, C. McDowell and R.D. Kratz of Cape Town; J. Truter of Riebeeck East; Victor Pringle of Bedford; E. Smith of Port Elizabeth; G. Smith of Pretoria; John Bell of Colesberg; and Reinhardt Kok of Nature's Valley.

My sincere thanks to the staff of Fernwood Press – publisher Pieter Struik, designer Wim Reinders, production coordinator Bunny Gallie, and Gaby Taube – and to Valerie Streak and Leni Martin for the final editing and popularizing of the text.

Lastly, I would like to pay tribute to John Winter, curator of Kirstenbosch National Botanical Garden, and to all concerned at the National Botanical Institute, especially Emsie du Plessis, Dr O.A. Leistner, Craig Hilton-Taylor and Dr H.F. Glen for the scientific editing, and Barrie Low, the director of education.

ERNST VAN JAARSVELD, MARCH 1994

THE SPONSOR

The publisher wishes to acknowledge, with thanks, the generous contribution made by First National Bank towards defraying the costs of this publication.

Frontispiece: *Gasteria pillansii* var. *ernesti-ruschii* (all parts x 1,5, except 1): 1. plant and inflorescence (x1); 2. perianth and stamens; 3. gynoecium; 4. cross section through mature leaf.

CONTENTS

FOREWORD

This volume on the genus *Gasteria* may be regarded as a companion volume to two others published within recent times on the related genus *Haworthia*. As all those interested in South African succulent plants will know, the first volume was by Bruce Bayer and the other was by Colonel C.L. Scott. The introductory information in these works corresponds closely to records applicable to *Gasteria* also.

My interest in the genus *Gasteria* dates back to 1925 and 1926 when I was in my early years of botanical survey in the eastern Cape, with headquarters in Grahamstown. There I had *Gasteria bicolor* var. *liliputana* and other species under close observation for many years, and still have in my possession type material of *G. baylissiana* Rauh, named after my good friends Roy and Estelle Bayliss. While I realized the desirability of a revision of the genus, at no time did I contemplate undertaking the task myself.

It is an indication of my long retirement from active botany that I have not met Ernst van Jaarsveld in person. I have nevertheless heard eulogistic comments from none other than my colleague, Dr L.E. Codd, who remarked particularly on Ernst's botanical enthusiasm and his 'eye for a species'. To me this is most refreshing at a time largely monopolized by computer science at the expense of essential personal taxonomy. Enthusiasm, of course, must be tempered by sound judgement and we are assured of this in Ernst. He has come to terms with an extensive synonymy. His conservative approach is reflected in that only one new name,* *Gasteria vlokii*, is established and that for a species isolated on the Waboomsberg in the Oudtshoorn district.

Subsequent to my commitment to this foreword I learned with pleasure that another associate, Ellaphie Ward-Hilhorst – already famous for her illustrations of species of *Pelargonium* and other South African plants – was to be responsible for the artwork for *Gasteria*. This collaboration ensures botanical fidelity and artistic presentation.

These circumstances confirm my first impression that the revision by Ernst van Jaarsveld will be an invaluable addition to South African botanical literature. It gives me pleasure to recommend it to the interested public.

R.A. DYER

Footnote from the author and the illustrator: *It is deeply regretted that Dr Dyer did not survive to see this publication. However, as a token of our great esteem and affection, we have decided to use this foreword which he wrote shortly before his death on 26 October 1987.*

**Four new* Gasteria *taxa have since been described:* Gasteria ellaphieae, G. glomerata, G. carinata *var.* retusa *and* G. brachyphylla *var.* bayeri.

Part 1

GASTERIA: AN INTRODUCTION

The popular succulent genus *Gasteria* is endemic to South Africa and has its main centres of distribution in the dry karroid and savanna regions of the south-eastern Cape. It belongs to the family Asphodelaceae, subfamily Alooideae (*sensu* Dahlgren *et al.* 1985; family Aloaceae *sensu* Cronquist 1981), which includes succulent plants with leaves arranged in a rosette or remaining opposite (rarely three- or five-ranked, but not in *Gasteria*), and with tubular flowers. Gasterias are at once distinguished from related genera by their recurved pendulous pedicels and belly-shaped (gasteriform), curved perianths, usually loosely arranged on the spreading racemes.

The plants come mainly from the eastern Cape, where the erratic rainfall pattern caused them to evolve as drought-resistant, shade-loving and shallow-rooted succulents. As a result of the climatic idiosyncracies of their natural habitat, they adapt well to indoor conditions, and have been popular among botanists and collectors in Europe since the late seventeenth century.

Nonetheless, *Gasteria* has been a problem genus ever since it was described by H.A. Duval in 1809. Early European botanists, who did not understand the considerable morphological variability that occurs in the species under natural conditions, described as 'new' numerous 'species' which, in reality, differed only in minor vegetative features. To add to the confusion, gasteria plants often flower in a juvenile stage, when the leaves may be markedly different from those of adult plants, and again this leads to erroneous 'discoveries' of new species. For example, *Gasteria nitida* was first described from a juvenile plant as *Haworthia nigricans*.

The extreme variability and chameleon-like nature of the species have in fact resulted in the publication of about 100 names in the course of the past three centuries. (In the last revision [Berger 1908], the number of recognized species was reduced from 85 to 43; then, between 1929 and 1940, the eminent amateur botanist, Karl von Poellnitz, described a further 25 'species'.) In most cases these names were extremely difficult to apply to the plants studied for this revision, as prior to the activities of Karl von Poellnitz, hardly any reference was made to the original habitat of the various named species. Early explorers and travellers avoided the arduous task of preparing dried specimens of *Gasteria* species, and indeed of most succulents (modern collectors are generally no different), preferring to collect seeds and live plants for introduction into the botanical gardens of Europe. This accounts for the paucity of specimens of succulent plants in herbaria today, while those rare, early dried specimens are fragmentary and of poor quality. Unlike most non-succulent plant species, which have been studied mainly from dried specimens, gasteria plants were illustrated and described from living plants growing in botanical gardens and private collections, without the preservation of a representative herbarium specimen.

Although succulents have held an increasing fascination for me since my student days, my interest in gasterias was only awakened in the late 1970s when I accepted a transfer from the Lowveld National Botanical Garden in Nelspruit to Kirstenbosch, as curator of the botanical garden's succulent collection. There I was regularly visited by the late Frank Stayner, curator of the Karoo National Botanical Garden, who often 'talked gasterias' to me. Probably because of the chaotic state of their classification, my interest was mild at first. However, it grew when, on the death of Frank Stayner, his notes, colour slides and photographs, as well as Berger's *Revision of* Gasteria, which had appeared in *Das Pflanzenreich* in 1908, passed into my hands.

Initially, I was interested principally in the collection and propagation of these plants for their horticultural value, but later I became intrigued by their variability – often within the same population group. As I studied the development of young gasterias it became clear that virtually all juvenile leaves differ markedly from those of the adult plants, and I soon realized that I was able to group certain species according to their juvenile epidermal features, enabling me to classify them below the generic level.

At this stage I enlisted the co-operation of Ellaphie Ward-Hilhorst, one of our foremost botanical artists, and she began the task of making watercolour illustrations of all the species in juvenile and adult flowering stages, where justified.

During the course of my investigations, which took place over a period of nine years, not only were the large living collections of *Gasteria* studied at the national botanical gardens at Kirstenbosch, Worcester and Pretoria, but live plants obtained from various European botanical gardens were also studied, as were dried specimens in South African and European herbaria. Field work took me to the habitat of every known species. Plants were collected and photographed, and soil and rock samples were taken at collecting sites to give me a clearer understanding of the growing requirements of the species. Herbarium records were prepared from cultivated plants and those collected in the field. All species were propagated from seed and from leaf cuttings (with the exception of *Gasteria rawlinsonii*, which was propagated from stem cuttings), and leaf development was observed and recorded.

My investigations have resulted in the naming of five new *Gasteria* taxa. The first, *Gasteria vlokii*, was found by Jan Vlok in 1984 on the Great Swartberg. This was the first gasteria to be found growing at an altitude of over 1 000 metres in a fynbos habitat with high rainfall. In 1991 I described *G. ellaphieae* and *G. glomerata*, both of which I had found almost accidentally while exploring the sheer rocky krantzes of the Kouga Dam near Hankey in the eastern Cape.

Gasteria carinata var. *retusa* was also described as new. A variety from the Worcester region, it was first collected in 1924 at Keerom Buttress by Hugo Naudé, the distinguished South African artist. This was published in 1992 in *The genus* Gasteria, *a synoptic review* in *Aloe*, journal of the South African Succulent Society. In the same issue, *G. brachyphylla* var. *bayeri*, collected near Ladismith in 1932, was described for the first time.

This revision, which recognizes only 16 species, aims to produce a simple, logical and natural classification. Taxonomic decisions have been based on the morphology and distribution of the species, which were delimited by discontinuities in two or more of their diagnostic features, while the arrangement of species reflects an evolutionary analysis which leads to a better understanding of the relationships between species.

For information on early collectors, explorers and botanists Gunn & Codd (1981) was consulted, as was Wijnands (1983). Vegetation units were described according to White (1983) and Acocks (1953). Rock and soil formations associated with the plants are described after Visser (1984).

Without written records it is impossible to say how long gasterias have fulfilled a role in the lives of the indigenous peoples of South Africa. But as soon as primordial man emerged from the African rain forests and began adapting to new environments, survival became dependent on his acquiring an intimate knowledge of his habitat: in particular its places of shelter and safety, its sources of water, its animals and its plants. Man became a hunter-gatherer, and in the course of his rovings and expansion into fresh territories in the southern portion of the continent he discovered the gasteria. Stone Age artefacts found in or near most gasteria habitats are strong evidence of this.

We know that gasterias were among the many plants utilized as a food by the Khoi-San, who prepared the flowers as vegetables. The Nguni, on the other hand, who came from the north and settled in the south-east, are recorded as having used the plants for medicinal and magical purposes.

When Europeans settled in southern Africa in the latter half of the seventeenth century, the Age of Enlightenment had already spread its wings, exciting and challenging the intellect, refuting the tenets of ignorance and superstition. The world was no longer regarded as flat, more widespread education and the availability of printed books had taken learning out of the monasteries, while the natural curiosity of man had been aroused by tales of brave new worlds beyond the sea – worlds with, among other things, wondrous people, animals, birds and plantlife. Man had fallen in love with knowledge, and the collecting of plants – among them gasterias – took place principally in response to his insatiable fascination for the natural world as a whole and his passion for the expanding science of botany in particular.

The interest in gasterias was nurtured not only by botanists, but also by collectors in Europe. The strange aloe-like plants and other succulents had an irresistible appeal for these men who, with their heated greenhouses, had opened up a new indoor horticultural activity. Adrian Haworth wrote of the popularity of these plants in his first revision of *Aloe* (1804): 'In the beginning and middle of the late century when the management of greenhouse and hothouse plants was in its infancy, when collections of exotics were neither so numerous nor so extensive as at present; not only succulents in general but those of the genus *Aloe* in particular were held in the highest repute both in Britain and on the continent, but especially more in Holland. That was the Golden Age of succulent plants.'

Botanical gardens in Europe were the initial and main driving force behind botanical exploration from the seventeenth to the nineteenth centuries. These gardens – and the subject of botany – were then closely linked to the medical profession, and were usually attached to universities. Their main objectives were not only to provide plants for the medical profession and to educate medical students, but also to open up a new and growing science, namely the classification of plants (Stearn 1961; Wijnands 1983).

During the seventeenth and the mid-eighteenth centuries, plants from remote parts of the world were mainly channelled to Europe through the Dutch botanical gardens, as the Dutch were the leading explorers of the time. However, for a number of reasons, during the latter half of the eighteenth century the Royal Botanic Gardens at Kew became the main source of supply, and it was mainly through the inspiration of the horticulturists there that botanical exploration on a methodical footing was ensured.

Organized botanical exploration within South Africa was inaugurated during the latter part of the nineteenth century with the establishment of universities, botanical gardens and other educational centres. It was Karl von Poellnitz, a keen German gasteria student, who stimulated interest in the species by his numerous articles, as well as correspondence with avid South African succulent growers and horticulturists. This interest inspired two South Africans, Frank Stayner (a previous curator of the Karoo National Botanical Garden) and Professor Ted Schelpe of the University of Cape Town, to undertake gasteria collecting expeditions in the late 1950s. An article written by Professor Schelpe in 1958 indicated that he planned to undertake a more comprehensive revision of *Gasteria*. However, his attention became drawn to orchids and ferns, to the detriment of his work on gasterias. Frank Stayner kept the gasteria flame alive and it was he who inspired the present author, resulting in this work.

In the seventeenth and eighteenth centuries the vegetation of the interior regions of southern Africa was little known. Plant collecting was undertaken by exploring parties who travelled in oxwagons, often under extremely difficult conditions. The demands of such an expedition are described by Stearn (1958): 'Botanical exploration calls for more than botanical knowledge. It calls for curiosity, zest, mental tenacity and a capacity to endure drudgery, hard work, discomfort and privation. It has to be encouraged intellectually, if not backed financially as well, by appreciative recipients of its hard-won material. Hence it may be said to reflect the mental climate of the time and the mental micro-climate of the individual. Moreover, the means of travel to and into a given region necessarily regulates the work that can be done there. If, then, the amount of botanical exploration up to the time of Linnaeus seems comparatively meagre, it should be balanced against the difficulties of travelling and the rarity of finding an eager, informed, tough individual with the desire and means to travel and the good fortune to return safely with his material, without being imprisoned, robbed, murdered, eaten or drowned, or enfeebled or slain by disease.'

THE SEVENTEENTH CENTURY

In 1682 a botanical garden (Medicyn-Hoff) specializing in medicinal plants and exotics was founded in Amsterdam and was soon to become one of the world's foremost botanical gardens. Jan Commelin, a spice merchant and pharmacist, and Joan Huydecoper, a leading and influential citizen who had been made a director of the Dutch East India Company in 1666 and served as mayor of Amsterdam from 1673 to 1693, were appointed as 'Commissarissen' (Wijnands 1983).

Some years earlier, in 1652, the Dutch East India Company had established a small settlement at the southern tip of Africa, its purpose being to provide fresh food for passing ships. Small wonder that the influential Huydecoper and Commelin, together with Caspar Commelin, the nephew of Jan Commelin and professor of botany at the Amsterdam Athenaeum, were responsible for the introduction to the Medicyn-Hoff of a large number of exotic plants, especially those of the Cape. Indeed, their ties with Simon van der Stel – governor at the Cape from 1679 to 1699 – and certain Dutch East India Company spice merchants resulted in the introduction of many new exotic plants to Amsterdam (Stearn 1961; Karsten 1951; Wijnands 1983).

The patronage and influence of these wealthy men made possible the employment of several skilled botanical artists who were responsible for a series of superb colour plates depicting the newly introduced plant species, some of which are reproduced in Commelin's *Horti medici amstelodamensis* (1697-1701) and later in Wijnands (1983). Several forms of *Gasteria disticha* and *G. carinata* were featured.

The Company's garden

The Dutch East India Company garden at the Cape was established soon after the settlement was founded in 1652. In 1679 Simon van der Stel, who had received some horticultural training in the Netherlands and who also had an interest in botany, was appointed governor and immediately set about making great improvements to the Company's garden. His keen interest in plants, as well as his ties with the Commelins, resulted in regular consignments of plants to the Medicyn-Hoff in Amsterdam. Realizing the wealth and value of the Cape flora, Van der Stel saw the need for a professional botanist, and in 1691

The earliest illustration of *Gasteria disticha,* by Jan Moninckx, was reproduced in Commelin's *Horti medici amstelodamensis* (1701). This is also one of the earliest *Gasteria* illustrations, and the iconotype of *G. disticha*.

Hendrik Oldenland was assigned to the post of master gardener of the Company's garden.

Oldenland, a Dane by origin who had trained at Leiden, came to the Cape in 1688 (Wijnands 1983). He was a physician and a skilled botanist, and the first to be appointed at the Cape. He soon started work on a 'kruid boek' or herbarium, in which each specimen was accompanied by a Latin phrase name, including 28 '*Aloe*' specimens which embraced *Gasteria* and *Haworthia* (Reynolds 1950). There are three phrase names in the 'kruid boek' representing two *Gasteria* species (*G. disticha* and two varieties of *G. carinata*), and it may be inferred that these were collected by Oldenland in 1696 during a three-month journey that he made to the eastern Cape, with a certain Isaq Schrijver, to buy cattle. Oldenland never published his list of species, but Valentyn (1726) did.

The party travelled past the site of present-day Caledon and along the southern foothills of the Langeberg to the site of present-day Heidelberg. From there they went over the Langeberg (Attaquasberg) and north-east to the Oudtshoorn district, then eastwards again to what is now Uniondale. Their course then took them northwards past Willowmore to the Aberdeen district and back. This route corresponds well with the *Gasteria* species described by Oldenland in his 'kruid boek', the area between Caledon and Heidelberg being the habitat of *G. carinata*. Judging from Commelin's (1701) illustration of *G. disticha*, it is probable that the plant was collected along the lower Breede River valley between present-day Bonnievale and Swellendam.

Two of the earliest gasteria paintings that exist have been attributed to either Heinrich Claudius or Oldenland. They are of *Gasteria pillansii* and *G. carinata* – both thought to be part of the documented historical collection of early Cape botanical and zoological paintings, *Codex Witsenii*. Heinrich Claudius (1655-1697) was a German naturalist and illustrator employed by the Dutch East India Company, by whom he was later sent to the Cape, where Simon van der Stel used him on several expeditions to illustrate natural history specimens. If Claudius was responsible for the two paintings in question, he probably illustrated *G. pillansii* in 1683 while accompanying Oloff Bergh on his Namaqualand expedition. *G. pillansii* is widespread in Namaqualand from north of Citrusdal and occurs on a well-known historical route.

Hendrik Oldenland died in 1697 and was succeeded by Jan Hartog, a horticulturist and plant collector from Leiden University, who had come to the Cape in 1691. Two years after Hartog had taken up his new post, Willem Adriaan van der Stel succeeded his father as governor of the Cape. He was a great plant lover and continued his father's work by sending consignments of plant specimens and seeds to the Medicyn-Hoff to augment its collection. Acting on instructions from Van der Stel to collect indigenous plants and seeds, Hartog undertook an expedition to the Riviersonderend Mountains between November 1699 and February 1700. Records show that while there he collected four '*Aloe*' species, which might well have included species of *Gasteria*.

EIGHTEENTH-CENTURY COLLECTORS AND CONTRIBUTORS

Carl Linnaeus (1707-1778), founder of our present plant and animal binomial system, was a Swede who received his doctorate in Holland. He later became a physician to George Clifford at Haarlem and eventually professor of medicine and botany at the University of Uppsala. In his *Species plantarum* (1753) he included the first gasteria under the name *Aloe disticha*. His circumscription, however, included four entities which today we know as *Gasteria disticha*, *G. carinata*, *G. carinata* var. *verrucosa* as well as *Aloe plicatilis*. It is rather amusing that Haworth (1827) considered this arrangement of Linnaeus to be 'Sad testimonies these, of the low state of that great man's knowledge in this department of Botany'.

Johann Auge (1711-1805), a botanical collector and horticulturist at Leiden University, came to the Cape in 1747 and was employed by Governor Hendrik Swellengrebel in the Company's garden, where he later became superintendent. Between 1761 and 1773 his journeys to collect plant specimens and seed took him to remote parts of the country, including Karasberg in southern Namibia as well as the eastern Cape. From 1772 to 1773 he also acted as guide to Carl Thunberg. Although there are no records that he collected any *Aloe* specimens, it is probable that Auge was responsible for introducing *Gasteria bicolor*, *G. acinacifolia* and *G. pulchra* into horticulture since these plants occur in the regions in which he travelled.

Phillip Miller (1691-1771), a well-known English horticulturist of the eighteenth century, published *Aloe carinata* and *A. verrucosa* in his *Gardener's dictionary* in 1768. Miller was employed at the Company of Apothecaries' botanical garden at Chelsea (Chelsea Physick Garden) in London from 1722 to 1770. He was an extremely energetic man who became very well known because of his publication, *The gardener's dictionary*, of which eight English editions appeared, the first in 1731 and the last in 1768, and which was the product of his wide range of practical experience. Through this monumental work he stimulated interest in the growing of a wide range of plants in England as well as in Europe. His book was so popular that translations into Dutch, German and French also appeared. In the eighth edition he adopted the Linnaean binomial classification system, in which he described *Gasteria carinata* var. *carinata* and *G. carinata* var. *verrucosa*, and *Aloe linguiforme*, a synonym of *G. disticha*.

Hendrik Swellengrebel (1734-1803), deacon of the Chapel of St. Martin's Cathedral in Utrecht, was another keen collector and grower of South African plants. The son of Hendrik Swellengrebel, governor at the Cape from 1739 to 1749, he spent most of his life in Holland. During a visit to the Cape between 1776 and 1777 he undertook three plant collecting expeditions. On his second journey into the interior of the eastern Cape in 1776, he collected *Gasteria nitida*, probably in the Langkloof or Humansdorp district, which species he later introduced to the botanical gardens in Utrecht.

Carl Thunberg (1743-1828), a Swedish botanist and a pupil of Linnaeus, is often referred to as 'the father of South African botany'. He arrived at the Cape in April 1772 and during his three-year visit he travelled extensively, collecting an astonishing 3 000 species. Among these were herbarium specimens of three *Gasteria* species – *G. bicolor*, *G. pulchra* and *G. carinata* var. *verrucosa*.

Thunberg went on three major inland journeys. On his second, from September 1773 to January 1774, he was accompanied by the Scottish horticulturist, Francis Masson, and we presume it was on this journey that he collected his three *Gasteria* species. Setting out from Cape Town, they travelled eastwards to the Outeniquas, along the Langkloof to somewhere near present-day Humansdorp and on to the Port Elizabeth district. *Gasteria carinata* var. *verrucosa* was probably collected in the Gouritz River valley. (Thunberg in fact named it *Aloe disticha*, but later N.E. Brown of the Royal Botanic Gardens at Kew considered it to be a new species and renamed it *G. thunbergii*; there is little doubt, however, that this was *G. carinata* var. *verrucosa*, previously described by Miller.)

Gasteria pulchra and *G. bicolor* would have been collected in the Humansdorp and Port Elizabeth districts where they both occur. The last-named species Thunberg called *Aloe maculata*, stating that he had collected it in the Outeniqua Mountains, which is not impossible since the species might well occur at the lower east end of that range. (The author has still to investigate the area.) Unfortunately, Thunberg's name could not be upheld because of Allioni's earlier *A. maculata* of 1774. Nor can the identity of the narrow-leaved (smooth) gasteria in Thunberg's herbarium be established with certainty since it is lacking in diagnostic detail. It is probably *G. pulchra* but it could also represent narrow-leaved forms of *G. bicolor*, or a hybrid between these species. Both species occur quite abundantly in the lower Gamtoos River valley, an area traversed by Thunberg and Masson.

Francis Masson (1741-1805) was a Scottish horticulturist who trained under William Aiton of the Royal Botanic Gardens, Kew. Through the influence of Sir Joseph Banks, Masson was sent on two expeditions to the Cape, first from 1772 to 1775, and later from 1785 to 1795. On both occasions he collected plants for the Royal Botanic Gardens at Kew. He was a very keen traveller and explorer, and an adventurous collector who discovered many new species. He also collected a few *Gasteria* species on his travels in the south-eastern region of South Africa, none of which was new.

NINETEENTH-CENTURY COLLECTORS AND CONTRIBUTORS

A.P. de Candolle (1778-1841) was a Swiss botanist at Geneva who produced a well-known work on succulent plants, illustrated by the renowned Pierre-Joseph Redouté (De Candolle 1799-1823). In this work four *Gasteria* species are treated, of which *G. pulchra* was described for the first time as a species under the name *Aloe obliqua*.

Adrian Haworth (1768-1833) was the leading authority on succulent plants in the first quarter of the nineteenth century. He was responsible for the first major revision of *Aloe*, which then included gasterias (1804). Haworth (1812) was also the first after Duval to acknowledge Duval's generic name *Gasteria* and later (1827) revise the genus *Gasteria*. A man of independent means, Haworth devoted his life to horticulture, botany and entomology (Stearn 1971). He was a keen grower of succulents, especially Aloineae and Mesembryanthemaceae. His articles on Aloineae appeared between 1804 and 1827 and he also corresponded with the highly regarded amateur botanist, Prince Salm-Reifferscheid-Dyck.

According to N.E. Brown, Haworth's plants were bought after his death by Wilson Saunders of Reigate, who passed them on to Thomas Cooper of that town. Cooper, in addition to being a noted Victorian collector of South African plants (commemorated in, among other names, *Aloe cooperi*), was Brown's father-in-law. In 1982 there were still a few 'Brown' plants – perhaps the last survivors of Haworth's collection – in the living collections at Kew.

Haworth is the author of *Gasteria bicolor*.

Henri Duval (1777-1814) was a Frenchman and a doctor by profession, who named the genus *Gasteria* in 1809. Very little else is known about him, except that he sent a copy of his publication to Haworth, who was the first to accept his genera *Gasteria* and *Haworthia*. Other well-known botanists of this period (Salm-Reifferscheid-Dyck, Aiton, Roemer and Schultes) did not accept this arrangement.

Baron Joseph von Jacquin (1727-1817), an adept scientist of Dutch origin, moved to Vienna in 1745 to further his studies in medicine and botany. Thereafter he became well known and respected as director of the Imperial Gardens at Schönbrunn, and it was to him that the botanical collectors Boos and Scholl sent their plants from the Cape. *Aloe acinacifolia* was described by Baron von Jacquin between 1811 and 1816 cited from a plant received from Holland and cultivated at Schönbrunn. It is not known who first introduced this species (*Gasteria acinacifolia*) to Europe; however, it seems very probable that it could have been either Swellengrebel or Auge. Unfortunately the holotype was destroyed during the Second World War, but the excellent illustration that accompanied Jacquin's description was selected by the author as an iconotype. (Baron von Jacquin is usually referred to as Jacquin and not as Von Jacquin.)

Prince Salm-Reifferscheid-Dyck (1773-1861) was a very keen amateur horticultural botanist and succulent enthusiast who built up one of the finest succulent collections of his era. He was very productive both in his numerous publications and in his outstanding botanical illustrations, having been privileged to have studied under the celebrated botanical artist Pierre-Joseph Redouté. His most important work, *Monographia generum Aloes et Mesembryanthemi* (1836-1842), consists of seven fascicules which include a collection of 33 gasteria lithographs, each accompanied by a Latin description. Salm-Reifferscheid-Dyck is the author of *Gasteria nitida* and *G. brachyphylla*.

James Bowie (1789-1869) joined the Royal Botanic Gardens at Kew in 1810 and first visited the Cape in 1816. Initially he confined his plant collecting expeditions to the southern parts. Then, in 1818, he explored as far east as Knysna, and in 1819 he visited Plettenberg Bay. His third extensive journey took him from Cape Town to the Langkloof, Humansdorp, Grahamstown and Port Elizabeth districts. On his last journey, however, he first sailed to Algoa Bay, and from there travelled inland as far as Graaff-Reinet. In 1823 he was recalled to Britain, but returned on his own account in 1827 to settle permanently in the Cape (Gunn & Codd 1981; Reynolds 1950).

Bowie introduced a number of *Gasteria* species to Kew, providing detailed habitat descriptions for five of them. These were published in *The South African Quarterly Journal* (1830). A collection of his drawings of Cape plants is housed at Kew.

Johann Drège (1794-1881) was a German horticulturist and botanical collector who arrived at the Cape in 1826. One of the greatest botanical explorers of his time, he travelled extensively throughout South Africa during his eight-year visit, and collected more than 200 000 specimens

Prince Salm-Reifferscheid-Dyck's illustration of *Gasteria nitida* var. *nitida* from his book *Monographia generum Aloes et Mesembryanthemi* (1836-1842). He was responsible for naming this species, and this plate was chosen as the iconotype for *G. nitida* var. *nitida*.

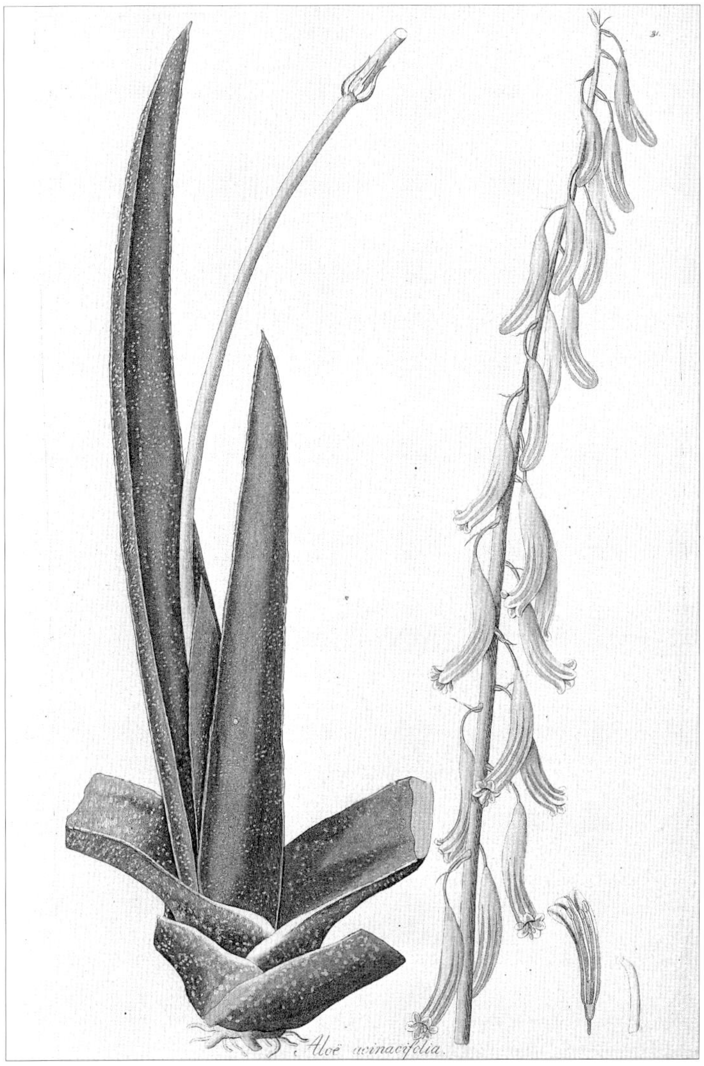

The earliest illustration of *Gasteria acinacifolia*, which appeared in Jacquin's *Ecologae plantarum rariorum* (1811-1816).

in the eastern and southern Cape. He recorded 25 *Aloe* species (1843), which included some *Gasteria* species. *G. maculata* var. *dregeana*, named for him by Berger (1908), in this revision is regarded as a synonym of *G. bicolor* (Gunn & Codd 1981; Reynolds 1950).

Christian Ecklon (1795-1868) and **Carl Zeyher** (1799-1858) were German botanists who travelled and worked together in South Africa. They collected some *Gasteria* specimens in the eastern Cape, and Salm-Reifferscheid-Dyck later named a species for Zeyher; however, it is undoubtedly a form of *G. bicolor*. Ecklon was responsible for introducing seeds of *G. acinacifolia* to Europe (Salm-Reifferscheid-Dyck 1854, cited 1836-1863).

Thomas Cooper (1815-1913) was a horticulturist and plant collector who was sent to the Cape between 1859 and 1862 by Wilson Saunders of Reigate. He introduced many Cape species to England, including a number of *Gasteria* species. It was Cooper who discovered and introduced *G. excelsa* to Kew. According to Berger (1908), he also introduced *G. carinata* var. *verrucosa*, *G. bicolor* and *G. acinacifolia*. (See also under Haworth.)

Sir Joseph Hooker (1817-1911), a British botanist and explorer, described *Aloe croucheri* (*Gasteria croucheri*) in 1869 from plants collected by Thomas Cooper. The description, which appeared in *Curtis's Botanical Magazine*, was accompanied by an excellent colour plate. Hooker, who was the son of a former director of the Royal Botanic Gardens at Kew, Sir William Hooker, was himself appointed director of the Gardens in 1865.

John Baker (1834-1920), British botanist and Keeper of the Herbarium at Kew, revised the genus *Gasteria* in 1880 and later also wrote it up for *Flora capensis* (1896). He is the author of 19 *Gasteria* species, of which only *G. excelsa* is recognized in the present revision. Baker described his *G. excelsa* in 1880 from plants collected by Thomas Cooper.

TWENTIETH-CENTURY COLLECTORS AND CONTRIBUTORS

Selmar Schönland (1860-1940), a botanist of German origin, settled in South Africa in 1889 where he was appointed curator of the Albany Museum. In 1905 he became the first professor of botany at Rhodes University, Grahamstown (Gunn & Codd 1981). Being in the centre of gasteria distribution, he collected a number of *Gasteria* specimens. He is the author of *G. armstrongii* which is treated in this revision as a variety of *G. nitida*.

Alwin Berger (1871-1931) was responsible for the last revision of *Gasteria* in 1908. He devoted most of his life to the taxonomy of succulent plants, gaining experience from various botanical gardens, including those of Dresden and Greifswald, Fratelli Rovelli's nursery at Pallanza in Italy, and the Palmgarten in Frankfurt. Berger later became curator of

Hortus Mortolensis, Sir Thomas Hanbury's famous garden at La Mortola, Italy, where he spent 19 years. It was here that his work on the Aloineae proceeded, which led to the last revision of *Gasteria* (1908). As a horticultural botanist he worked mainly from living specimens (Herre 1971), although the Botanical Museum in Berlin houses a collection of his dried specimens.

Neville Pillans (1884-1964) was a botanist at the Bolus Herbarium of the University of Cape Town when he collected *Gasteria pillansii* in the Clanwilliam district. He was an active collector and explorer, and had an outstanding succulent collection at his Cape Town home (Gunn & Codd 1981; Reynolds 1950).

Louisa Bolus (née Kensit) (1877-1970) was one of South Africa's most productive botanists. She probably described more new species than anyone else in South Africa, although most of these were based on material sent to her by correspondents. Her own field work seems to have been restricted to relatively few expeditions to the Clanwilliam area and to Namaqualand. She became curator of her uncle's (Harry Bolus) private herbarium in 1911. This herbarium was later donated to the University of Cape Town where she remained curator until her retirement in 1955. She is the author of *Gasteria pillansii*.

Gilbert Reynolds (1895-1967) was an optometrist by profession and became interested in aloes in 1930. He travelled widely in southern Africa, and later also in east and north Africa and Madagascar, collecting aloe plants. He was responsible for producing two major works on aloes (1950, 1966). He collected a number of *Gasteria* species in the eastern Cape; these are housed in the Bolus Herbarium of the University of Cape Town.

Karl von Poellnitz (1896-1945), a German scientist who studied agronomy and natural science, was a dedicated succulent student, describing 25 'new' *Gasteria* species between 1929 and 1940. His most important gasteria publication, *Zur Kenntnis der Gattung* Gasteria (1938a), was in many ways the foundation of this revision of the genus. He corresponded with a number of South African horticulturists and botanists, including Stayner, Long and Herre, thus stimulating new interest in gasterias. Von Poellnitz died tragically in a bombing raid towards the end of the Second World War. Photographs of his plants are housed in the botanical garden at Dahlem, Berlin. Jacobsen (1960) named *G. poellnitziana* in his honour; in this revision the name is regarded as a synonym of *G. pulchra*.

Herman Jacobsen (1897-1978), a German horticulturist, started his career as a succulent specialist in 1926. In 1929 he was employed at the botanical gardens in Kiel and remained there until his retirement. His three-volume work on the succulent plants of the world, in which 75 *Gasteria* species are treated, was published in 1954 and still remains a standard work on succulent plants.

Hans Herre (1895-1979), a German horticulturist trained at Woerlitz, Germany, arrived in South Africa in 1925. He was appointed curator of the botanical gardens at the University of Stellenbosch, where he remained until his retirement in 1963. His main interest was succulent plants, especially Mesembryanthemaceae. He was honoured in *Gasteria herreana* by Von Poellnitz, but in this revision it is regarded as a form of *G. bicolor*.

Roy Bayliss (1909-1993), a botanical explorer, travelled extensively in eastern and southern Africa. He was a keen and active collector of succulent plants and discovered many new species. *Gasteria baylissiana*, which he collected together with the discoverer Mr John Truter in the Suurberg Mountains, was named after him. He settled in Grahamstown, from where he collected most of his *Gasteria* species, but later moved to Zimbabwe.

Werner Rauh (1913-) was professor of botany at the University of Heidelberg, Germany. His main interest is succulent plants and he has visited South Africa several times. Professor Rauh is the author of *Gasteria baylissiana*.

Allen Dyer (1900-1988) joined the staff of the Botanical Research Institute in 1925 and was stationed in Grahamstown where he later became curator of the Albany Museum herbarium. He was subsequently transferred to the National Herbarium in Pretoria where he served as director from 1944 to 1963. Having a keen interest in succulent plants, he was involved with their taxonomy, especially Asclepiadaceae and Euphorbiaceae. He collected a number of *Gasteria* specimens, of which *G. bicolor* var. *liliputana* (which he collected together with Miss Lilian Britten) was new.

Frank Long (1884-1961) was an English horticulturist trained at Kew, who later joined the Port Elizabeth Parks Department. He was a keen succulent collector in the eastern and central Cape. *Gasteria longiana* was named after him. In the present revision it is regarded as a form of *G. bicolor*.

Amelia Mauve (née Obermeyer) (1907-) was curator of the Transvaal Museum herbarium and became a member of the Botanical Research Institute in 1957. Her special interest was in Monocotyledonae. She collected a number of *Gasteria* species in the southern and eastern Cape and is also the author of *G. rawlinsonii*.

Ellaphie Ward-Hilhorst (1920-) is well known for her illustrations of species of *Pelargonium*, which appear in three volumes of *Pelargoniums of southern Africa* by Van der Walt (1977) and by Vorster (1981, 1988), and in *A revision of the genus Haemanthus* by D. Snijman (1984). Her work has appeared in numerous issues of *The Flowering Plants of Africa*. Her illustrations of gasterias appear in this volume.

Gordon Rowley (1921-) is a well-known succulent authority and the author of several works on succulents. A lecturer in horticultural botany at Reading University, England, he is the author of *Gasteria batesiana*.

Frank Stayner (1907-1981) succeeded Jacques Thudichum in 1959 to become the third curator of the Karoo National Botanical Garden at Worcester, a position he held until his retirement in 1969. He was keenly interested in gasterias and corresponded with Von Poellnitz, who named *Gasteria stayneri* (a form of *G. nitida*) after him. It was Stayner who inspired the author to undertake this revision.

Edmund André Charles Louis Eloi (Ted) Schelpe (1924-1985) was appointed curator of the Bolus Herbarium at the University of Cape Town in 1955. His interest in gasterias resulted in an enlightening publication on the subject, and he was the first botanist to recognize that *Gasteria armstrongii* is no more than a form of *G. nitida*. Some of his living plants are still to be seen in the collections at Kirstenbosch.

Jan Vlok (1957-) is one of the foremost botanists of the fynbos and karoo regions in the southern Cape. He is a very active plant collector and has made a significant contribution to our knowledge of plants and their ecology in those areas, as well as having discovered many plant species new to science. He has collected a number of *Gasteria* species, of which *G. vlokii* was described in 1987.

GENERIC DELIMITATION

Gasterias belong to the subfamily Alooideae of the family Asphodelaceae (Dahlgren *et al.* 1985) which includes approximately eight genera: *Aloe* L. (330 species, Reynolds 1966, 1982), *Haworthia* Duv. (68 species, Bayer 1982), *Gasteria* Duv. (16 species), *Astroloba* Uitewaal (7 species, Reynecke 1965), *Poellnitzia* Uitewaal (1 species), *Chamaealoe* Berger (1 species), *Chortolirion* Berger (1 species, Smith 1985, 1991), and *Lomatophyllum* Willd. (13 species, Jacobsen 1974). Some workers, however, prefer to include *Chamaealoe* and *Lomatophyllum* in *Aloe*, and *Chortolirion* in *Haworthia*. The subfamily Alooideae is clearly monophyletic and has a common karyotype, consisting of four long and three short chromosomes (Riley & Madjumdar 1979). The genus *Aloe* (earliest described of the Alooideae) was described by Linnaeus in 1753. Berger reviewed the tribe Aloineae (Liliaceae then comprising 28 tribes) in 1908 in which he recognized two subtribes, Kniphofiinae and Aloineae. Hutchinson raised these to tribal status in 1934. Rowley, by means of numerical analysis, reviewed the generic status of the Aloineae in 1967. It was he who produced the first dendrogram or phenetic tree, depicting the relationships of the 13 genera. As he found the generic delimitation of most genera of the Aloineae to be unsatisfactory, he suggested further investigation. In a later article (1976a) he constructed a 'phylogeneric tree' depicting a historic review of the delimitation of the genera of the Alooideae. Recently (1982) the old and large heterogeneous family Liliaceae was restructured by Dahlgren & Clifford, and this is now widely accepted. It is divided into homogeneous natural groups, each representing a family of its own, of which Asphodelaceae is part.

The subfamily Alooideae is delimited by floral and vegetative characters; it is recognized by its rosette of succulent, usually spiny, leaves (distichous, spirally arranged or rosulate) which gradually, with age, wither from below. The racemose or paniculate inflorescence appears to be axillary but is in fact terminal, bearing fleshy tubular flowers (Dahlgren *et al.* 1985). It also differs from the subfamily Asphodeloideae by the vascular bundles, which are arranged in a ring around the central ground parenchyma and have caps of aloine cells at the phloem poles. These characters seemed to Cronquist (1981) to set this group so far apart from others as to make it fall into a separate family, the Aloaceae. In *Gasteria* and *Haworthia* these aloine cells are often replaced by sclerenchyma (Dahlgren *et al.* 1985).

The Alooideae are predominantly confined to Africa, Madagascar and adjacent regions, but the main centres of species diversity are in southern Africa. Apart from *Gasteria*, *Haworthia*, *Poellnitzia* and *Astroloba*, which are well circumscribed, the delimitation of the remaining genera is not clear, especially that of the large and heterogeneous *Aloe* (Rowley 1976a).

Gasteria is a well-defined, natural and monophyletic group within the Alooideae, which is clearly distinguished both in floral and in vegetative features, but the most significant diagnostic characters are found in the flowers. *Gasteria* is at once distinguished from related genera, especially *Aloe* to which it bears a close resemblance, by its lax and usually secundly arranged, recurved, pendulous pedicels and curved, basally gasteriform perianth, which only becomes erect after fertilization. All the other members of the Alooideae have spreading to erect pedicels bearing erect to pendulous flowers.

Important diagnostic characters, found in all gasteria flowers, are the membranous margins of the three inner perianth segments, which are free and channelled for the length of the gasteriform portion, and through which the nectar is projected. The maculate aloes also have curved, basally globose ventricose flowers superficially similar to those of *Gasteria*, but this should not be interpreted as indicative of close affinity with the former. The inner perianth segments of these aloes are fully fused to the outer segments in a globose, bulbous base much smaller than the gasteriform part of any gasteria flower; also, no member of this group has the tricoloured flowers characteristic of *Gasteria*. The similar bulbous nature of the flowers in these genera could be due to parallel evolution.

The flowers in *Aloe* are borne on erect or spreading, dense or lax, capitate to cylindrical racemes. Species of *Aloe* with lax racemes do not often have secund flowers, and those that do, bear them on dense racemes – certainly this is the case in South Africa.

Except for *Gasteria rawlinsonii*, all *Gasteria* species can be readily defined by their green, brittle, mottled, linear-lanceolate to lorate, succulent leaves which are distinctly keeled when arranged spirally and are not bitter to the taste. *Gasteria* differs also in its leaf margin, which is tuberculate. The semi-translucent cartilaginous tubercles merge towards the apex of the leaves, becoming entire, crenulate or serrulate. The leaves of *Aloe* species are tough, very bitter and usually armed with distinct prickles and do not proliferate from broken leaf pieces, as do those of *Gasteria*.

INFRAGENERIC CLASSIFICATION

The basic unit of classification is the species. According to Radford (1986), 'a species as a biologic, genetic and evolutionary product of speciation can be considered a basic, discrete biological unit with a distinctive set of correlated characters that are fixed for a moment in evolutionary time'. The species concept is far from clear-cut and can be regarded as an evolutionary product of human reasoning. Species concepts in the past have tended to fluctuate considerably. Linnaeus's species concepts of the eighteenth century were very broad (classification artificial), followed by a very narrow notion (taxonomic species) in the nineteenth and early twentieth centuries. The modern taxonomic multidisciplinary approach has had a more stabilizing effect on our concept of species. Our contemporary expanding knowledge of natural variation and distribution patterns has resulted in a broader species concept.

My task, in the light of previous *Gasteria* research, was to study gasteria populations throughout their known distribution range. A primary task was to identify variation (structural, developmental and functional), distinguish between phenotypic and genotypic variation, and then select characters (and their states) that remained constant or fairly constant and which could therefore be used effectively in communicating differences between taxa. The next step was the identification, circumscription and interpretation of discrete discontinuous units (taxa) objectively, arranging them in a hierarchy (for a phylogenetic classification system), and determining their interrelationships. To communicate these taxa effectively, the earliest applicable names had to be selected, and where no names existed, species names had to be given. The methods of achieving this are described below.

Genetically and environmentally controlled variation

The variation, both phenotypic and genotypic, in *Gasteria* is remarkable, and this only becomes apparent through studying plants in the field, and growing plants in the botanical garden, the latter being necessary to look at the juvenile phase (ontogeny), genetic variance and variation due to environment (phenotypic plasticity). This variability, in earlier studies, is reflected by the long list of synonyms, of which Schelpe (1958) had the following to say: 'The multiplicity of published species of *Gasteria* is ascribed to the ease of introduction of these plants to gardens in Europe and to the lack of appreciation of the variability of these plants in the field by the authors who described them.' Schelpe's approach to studying variability in the field was presented in his excellent article (1958) and I followed this up. He pointed out the phenotypic plasticity of the *G. carinata*, *G. nitida*, *G. pillansii* and *G. bicolor* complexes, but also, unfortunately, correlated almost all the variation he observed to environmental conditions.

Over an eight-year period at Kirstenbosch, I had the opportunity to study a clinal range of all species, from various parts in the distribution

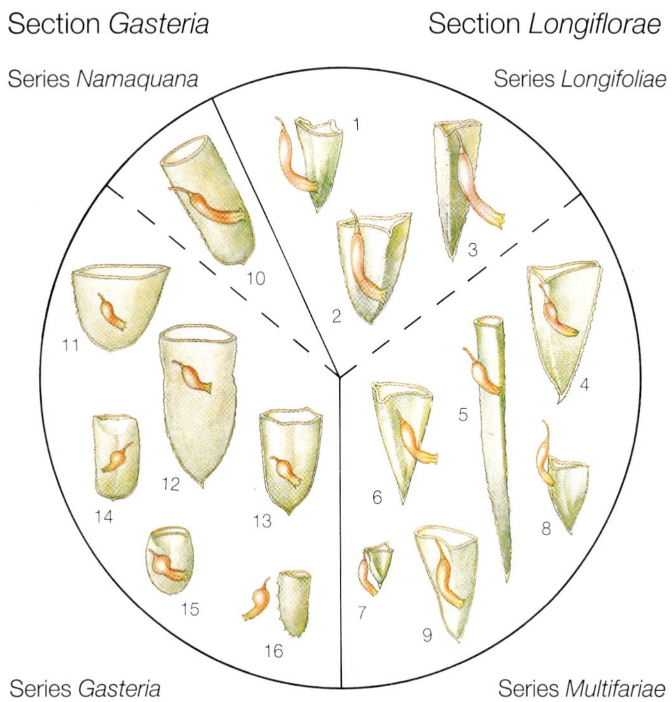

1 *G. batesiana*
2 *G. croucheri*
3 *G. acinacifolia*
4 *G. excelsa*
5 *G. pulchra*
6 *G. nitida*
7 *G. ellaphieae*
8 *G. vlokii*
9 *G. carinata*
10 *G. pillansii*
11 *G. disticha*
12 *G. brachyphylla*
13 *G. bicolor*
14 *G. baylissiana*
15 *G. glomerata*
16 *G. rawlinsonii*

A schematic synopsis of *Gasteria* classification. To the right of the solid line section *Longiflorae* is divided into series *Longifoliae* and series *Multifariae*. The second section, *Gasteria*, comprises series *Namaquana* and series *Gasteria*.

range, under identical cultural conditions. In cultivation I found that plants often retain differences that were initially thought to be due only to phenotypic plasticity and which are therefore the result of genotypic variation. Genotypic variation was also observed within small populations. The leaves of *Gasteria carinata* var. *verrucosa* at Infanta (Breede River mouth) and *G. bicolor* var. *liliputana* (Drivers Bush, east of Grahamstown) were found to vary considerably quantitatively. Grown under similar conditions, the difference in leaf length remained the same. Similarly, the variation in phyllotaxy of *G. bicolor* from Swartwaterspoort was correlated with environmental conditions and under genetic control. Plants with distichous and spirally distichous leaf arrangements were grown under similar environmental conditions: some of these remained distichous and others became spirally arranged. Unfortunately, Schelpe never continued his own preliminary investigation into the genus *Gasteria* as he turned his attention to Pteridophyta and Orchidaceae.

The designation of ranks

In the field, taxa were observed in as many different localities as possible and clinal variations were recorded. Continuous or gradual variation, usually in association with a similar habitat, with distinct morphological characteristics in common, was treated as a species. Where morphological variation between populations was discontinuous or abrupt, usually in association with a change in habitat, decisions on the rank of taxa were made.

Species arrangement and sectional and series ranks

Species arrangement and the identification of sectional and series ranks follow a cladistic analysis. My aim was to produce a natural classification reflecting the genealogical history and evolutionary relationships. It includes the accommodation or ranking of taxa in monophyletic groups. Two sections were identified. Differences here were based on the shape of the perianth, correlated with leaf shape and texture. Distinct differences in perianth shapes within the two sections warranted the subsectional rank of series.

The two sections, *Longiflorae* and *Gasteria*, are defined by floral and leaf shape, as well as by leaf epidermal texture. Section *Longiflorae* has features regarded as the more primitive (plesiomorphous). It includes the two series, *Longifoliae* (three species) and *Multifariae* (six species), and it incorporates species with flowers lacking the very conspicuous inflated perianth base so characteristic of section *Gasteria*. The leaves are triangular to triangular lanceolate and smooth or tuberculate. Section *Gasteria*, characterized by its lorate leaves and globose or globose-elliptic perianth base, is divided into two series, *Namaquana* and *Gasteria*. Series *Namaquana* contains the single species *Gasteria pillansii*, whereas series *Gasteria* contains six species. *Namaquana* is differentiated by the gasteriform portion of the perianth being less than a third of its length. Furthermore, the margins of the inner perianth segments are constricted below the apex of the ovary. The flowers of series *Gasteria* are gasteriform for more than half of the perianth length.

As this book is intended to help the widest possible range of users to identify their plants, sectional taxonomy is not discussed further here. Interested academic readers are referred to an earlier publication (Van Jaarsveld 1992).

Species rank

Species rank was assigned where variation was found to be abrupt (associated with two or more stable discontinuous characters) and usually correlated with a change in habitat. Sixteen species were identified. Clearly defined species include: *Gasteria pillansii*, *G. baylissiana*, *G. glomerata* and *G. rawlinsonii* (section *Gasteria*). The remaining three species (of the same section), *G. disticha*, *G. brachyphylla* and *G. bicolor*, were very closely related with less distinct species margins. Of section *Longiflorae*, *G. vlokii*, *G. ellaphieae*, *G. nitida*, *G. excelsa* and *G. pulchra* were clearly defined. The species boundaries of the remainder of section *Longiflorae*, *G. batesiana*, *G. croucheri* and *G. acinacifolia*, were less distinct. The variable *G. carinata* is placed in section *Longiflorae*.

Varietal rank

Extreme quantitative variation and minor morphological variation within the natural range of the species warranted the imposition of varietal rank in the cases of *Gasteria pillansii* var. *ernesti-ruschii*, *G. bicolor* var. *liliputana* and *G. brachyphylla* var. *bayeri*. *G. carinata* (section *Longiflorae*) from the southern Cape south of the Langeberg and Riviersonderend mountain ranges presented the most problems. Both the leaf phyllotaxy and epidermal texture varied continuously over a wide range and no abrupt clear-cut divisions could be established (upholding the older taxa such as *G. verrucosa*). The habitats of these variants, however, remained more or less constant.

The plants usually occur on shale outcrops in a mosaic of relic patches of renosterveld, valley bushveld, noorsveld, succulent karoo and occasionally limestone fynbos. Here the variation is thought to be a result of introgression (hybrid origin) from *Gasteria disticha* and *G. brachyphylla* (section *Gasteria*), which occur to the north of the Langeberg range. The reason for this is that three large river systems (the Breede, Gouritz and Brak) drain through the distribution range of *G. carinata*, and it is in the vicinity of these river systems that the greatest genotypic variation has been observed. The Gouritz and Brak rivers drain through the southern parts of the Great and Little Karoo (habitat of *G. disticha* and *G. brachyphylla*) before they enter the range of *G. carinata*, while the Breede River first drains through the Robertson Karoo (habitat of *G. disticha*). Flash flooding of these rivers is a regular phenomenon and plants of *G. disticha* and *G. brachyphylla* carried downstream could very easily become established on the banks of the

Gouritz or Brak rivers in the vicinity of *G. carinata* populations, resulting initially in hybridization (flowering times are similar) and, at a later stage, in speciation.

Gasteria carinata var. *verrucosa* has an asperulous epidermal texture similar to that of *G. disticha*. The status of *G. carinata* var. *retusa* is not fully understood. It occurs on the lower slopes of the Langeberg near Worcester and Robertson (close to *G. disticha* populations) and has features intermediate between *G. disticha* (section *Gasteria*) and *G. carinata* (section *Longiflorae*). In this revision it has been placed here tentatively as a variety of *G. carinata*.

Morphology

Gasterias are slow-growing xerophytic perennial plants which vary in height from the dwarf *Gasteria bicolor* var. *liliputana* (20 mm) to the large *G. acinacifolia* (600 mm). Most species shoot prolifically from the base and may form small to large dense groups, although *G. excelsa*, *G. acinacifolia* and forms of *G. nitida* may be solitary. *G. pillansii* proliferates from subterranean stolons, apparently a derived condition resulting in dense groups of up to 2 m in diameter. The habit of most species is decumbent to erect and rarely pendulous (*G. rawlinsonii*). Seeds of all the species sown in summer and autumn in the Kirstenbosch shade house germinated within 10 to 18 days. Compared to most aloes, gasteria seedlings grow more slowly and the general height of the seedlings after six months was approximately 10 mm, which is also slower than plants grown vegetatively from leaf cuttings. Plants of *G. baylissiana* grown from leaf cuttings in December 1986 flowered in October 1989.

ROOTS
Most species have a very shallow, spreading, adventitious root system, inhabiting the loose upper humus and leaf litter layer of the soil. The roots are slightly to fully succulent, terete and with a single layer of water-absorbing tissue (velamen) (Dahlgren & Clifford 1982). *Gasteria nitida*, *G. ellaphieae* and *G. vlokii* have roots which penetrate deeper, and are linear obclavate at the base, gradually dilating from the tips downwards.

STEMS
Almost all plants are acaulescent or very short-stemmed, except *Gasteria bicolor* and *G. rawlinsonii*. *G. bicolor* has a short decumbent to erect leafy stem which is of taxonomic importance. Initially the shoot growth is terminal, becoming sympodial from the first flowering (Berger 1908; Dahlgren *et al.* 1985). *G. rawlinsonii* is unique in having long, pendulous leafy stems up to 1 m in length, but because of the amplexicaul leaf bases, the stems are not visible. Due to the withering of lower leaves in old plants, the stem bases of *G. rawlinsonii* may become exposed, showing the cylindrical internodes.

LEAVES
The leaves of *Gasteria* species are distichous, spirally distichous or rosulate. The juvenile leaves, as found in other members of the subfamily Alooidae, are distichous at first, and in eight taxa this may persist. Most species (11), however, have leaves that soon become spirally arranged, either clockwise or anticlockwise, or densely rosulate. The leaves are more laxly arranged in the caulescent *G. bicolor* and *G. rawlinsonii*. The dorsiventrally flattened leaves are approximately two to six times wider than they are deep, with a firm leathery texture, but they are brittle, which is of generic importance. All species except *G. glomerata* have leaf spots, often arranged in transverse bands, but forms with an amalgamation of the spots have been observed in field populations of *G. pillansii* near Clanwilliam. The ontogeny of the leaves, their epidermal texture, shape and size are of taxonomic importance, but the phenotypic plasticity exhibited by most species should also be taken into account.

In almost all the species, the juvenile leaves differ markedly from fully grown ones, and in some cases plants may flower in the juvenile phase, which has caused confusion in the past (Schelpe 1958). The juvenile leaves of all the species are lorate, erectly spreading to patent, without a keel. The apex is obtuse or truncate with a distinct mucro. In cross section the juvenile leaves are biconvex, or with an upper (adaxially) flat surface, and a lower (abaxially) convex surface. The epidermal texture may be asperulous with larger tubercles, asperulous with no tubercles or, rarely, smooth. In section *Longiflorae* the juvenile leaves are, apart from being asperulous, distinctly tuberculate or, rarely, asperulous with no tubercles, but they become smooth and change their shape, both in outline and in cross section, in the adult stage. *Gasteria bicolor* and *G. brachyphylla* have asperulous juvenile leaves which become smooth with maturity. The leaf surface texture of juvenile and adult plants of *G. pillansii*, *G. disticha*, *G. vlokii* and *G. rawlinsonii* remains unchanged.

The adult leaf shape varies from lorate to triangular, rarely linear triangular or ensiform. Leaves that become spirally arranged usually have a distinct marginiform excentric keel. The upper surface varies from flat to rarely convex, but becomes deeply channelled and reddish coloured during drought. The lower surface is usually convex. The leaf apex also varies: it may be rounded, acute, obtuse or truncate. The margin, which is rarely entire, and keel are usually armed with cartilaginous, semi-translucent white tubercles, which often become entire towards the apex, forming a hard, semi-translucent margin which may be serrulate, crenulate or denticulate. *Gasteria excelsa* has a very sharp serrulate leaf margin. Leaves of some populations of *G. rawlinsonii* differ markedly

The juvenile and adult leaves of most *Gasteria* species show marked differences, especially those of *G. nitida* var. *nitida*. In this case the juvenile leaves are tuberculate and strap-shaped, whereas the adult leaves have a smooth, shiny surface and are triangular-lanceolate in shape.

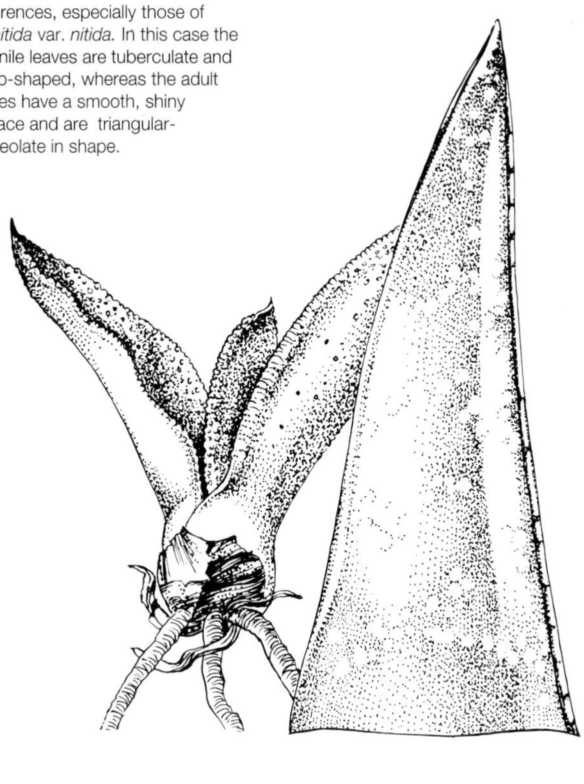

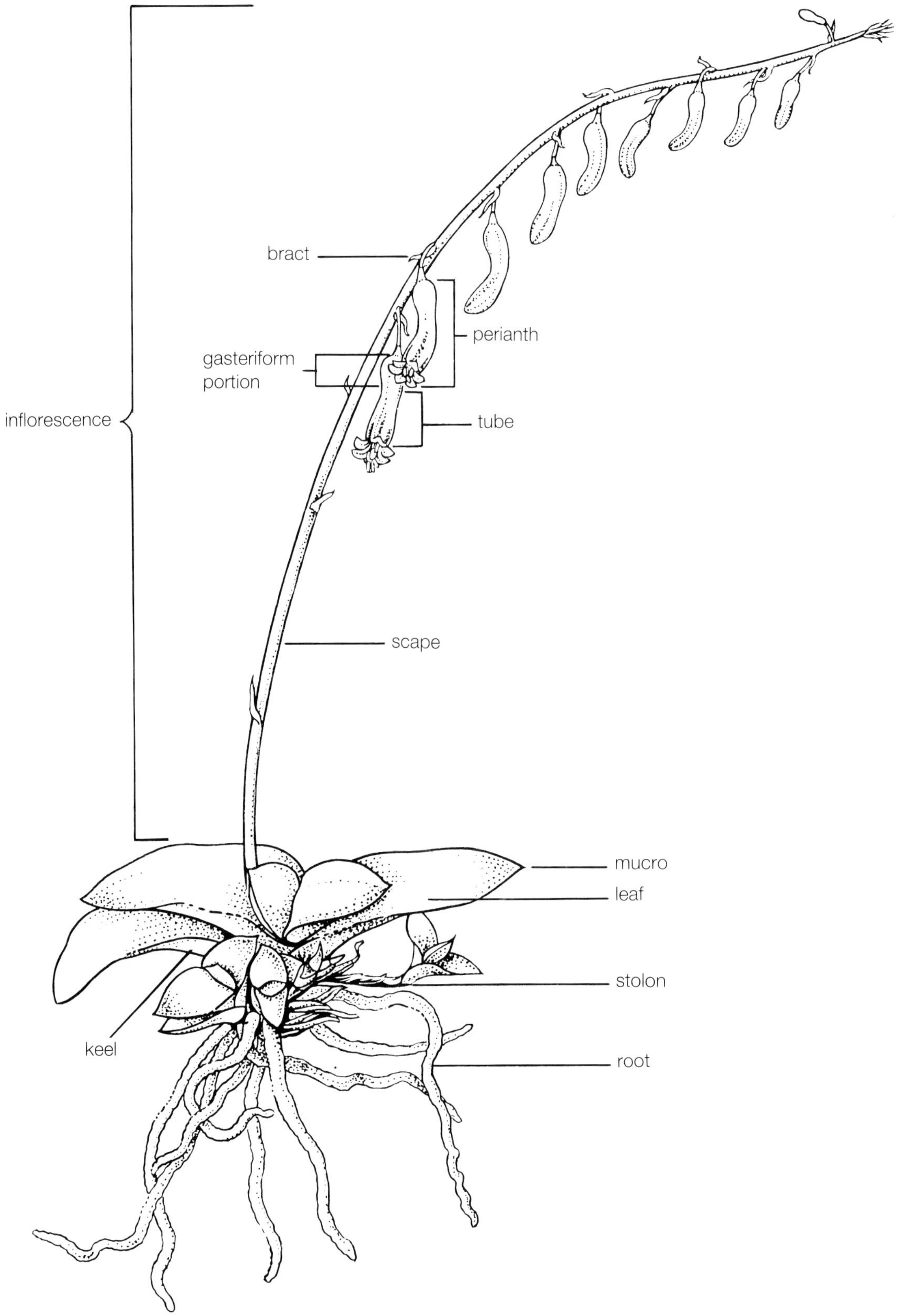

A mature *Gasteria* plant (*G. pillansii* var. *ernesti-ruschii*).

from those of any other gasteria because they are sparingly denticulate, with tubercles that may become black with age. The leaves of *G. bicolor* are twisted sideways. The leaf apex may vary from acuminate through acute and obtuse to retuse, and is usually armed with a distinct mucro which may be slightly excentric. The leaf apex in the adult form of *G. bicolor* is not symmetric in outline, the obtuse apex being oblique with an excentric mucro. Unlike that of the aloe, gasteria leaf sap is colourless and not bitter.

Gasteria species have a chameleon-like nature and adjust their leaves according to the availability of light. This phenomenon has been observed in most species in cultivation as well as in the field. A form of *G. carinata* in the De Hoop Nature Reserve has been observed growing partly in the shade of a group of *Rhus glauca*. The plants growing in shade were much larger with darker green, spirally arranged and sparsely tubercled leaves, while those growing in full sun were smaller, more compact and densely tubercled. This phenomenon was also observed in another species by Schelpe (1958).

Within a particular species, leaves may vary considerably in quantitative and qualitative features. One of the most variable and misunderstood species is *Gasteria carinata* of the southern Cape. Populations vary continuously and the leaves may remain distichous or become spirally arranged. At the western end of the distribution range the leaves are distinctly tuberculate, but become smooth towards the eastern end. Apart from the texture of the epidermis and the phyllotaxy, the size and leaf margin also vary.

INFLORESCENCE

An indeterminate inflorescence (one to three per plant) is borne terminally but appears superficially to be axillary, because it is pushed aside by the leaves in a pseudolateral position (Berger 1908; Dahlgren *et al.* 1985). The shoot thus continues its growth sympodially. The inflorescence in *Gasteria*, as in other members of the subfamily Alooideae, is racemose, or branched to form lax to dense panicles one to 12 times longer than the height of the plant. The shape and size are of taxonomic value. The majority of species have simple, erectly spreading, curved racemes that occasionally may have two side branches arising from near the centre of the inflorescence. The direction of growth in the southern hemisphere is usually northwards, towards the light source. The inflorescences of *G. croucheri*, *G. acinacifolia*, *G. excelsa*, *G. nitida* var. *nitida*, *G. bicolor*, *G. disticha*, *G. brachyphylla*, *G. baylissiana* and *G. pulchra* are branched; the erectly spreading side branches are arranged alternately on the central axis or main branch, which is usually longer than the side branches. The larger species may have side branches that have secondary branching. The lax panicles of *G. acinacifolia* and *G. croucheri* are flat-topped due to the side branches and main axis being horizontally curved.

The side branches of the inflorescences of *Gasteria croucheri* from Oribi Gorge are often drooping. The bases of the scapes are flattened and the sides acute, gradually becoming terete above. The scape of *G. pillansii* is minutely and sparsely asperulous. *G. excelsa* has a flattened scape, convex on one side. The lower bracts are ovate to broadly ovate,

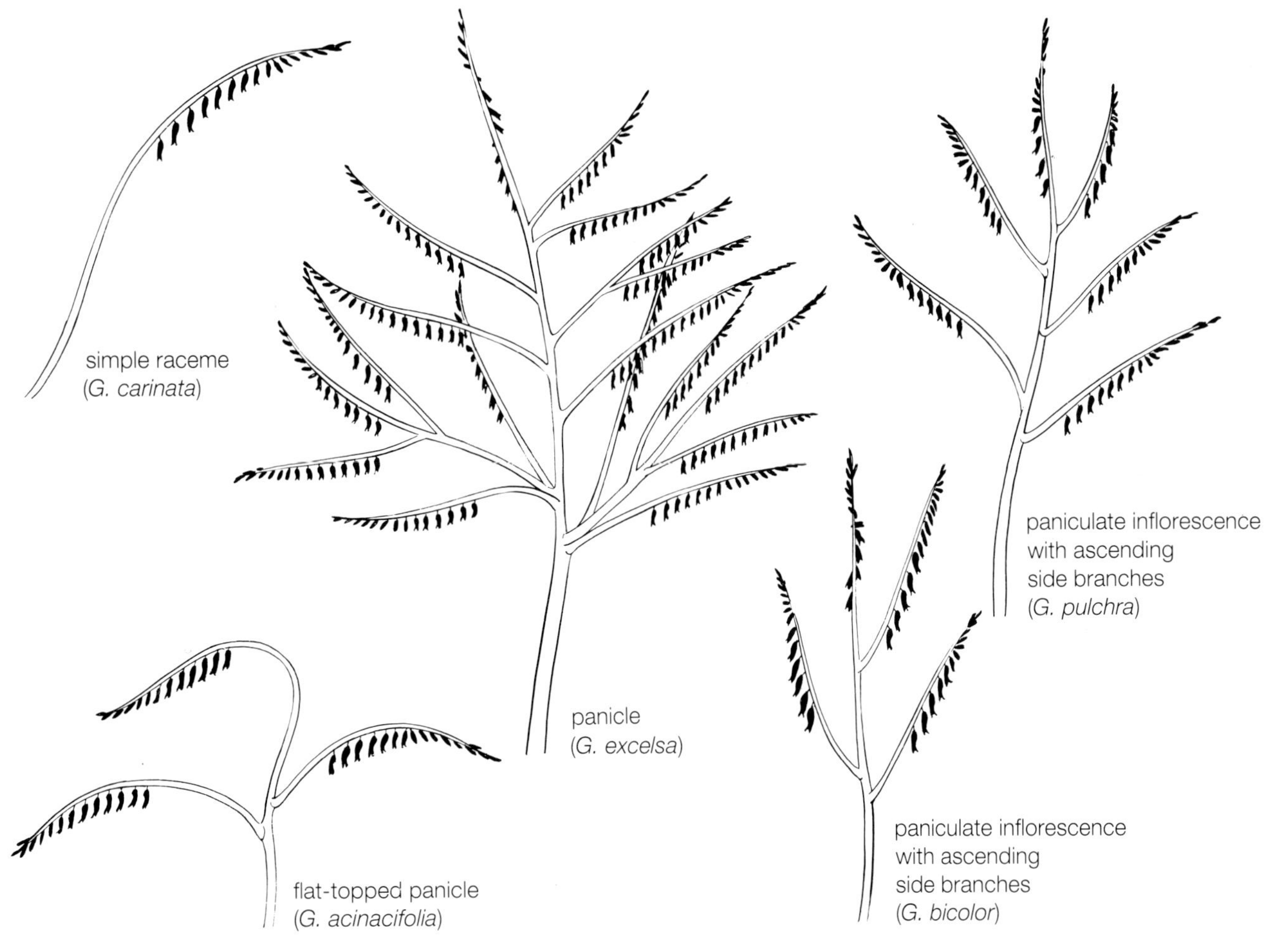

Gasteria inflorescence types.

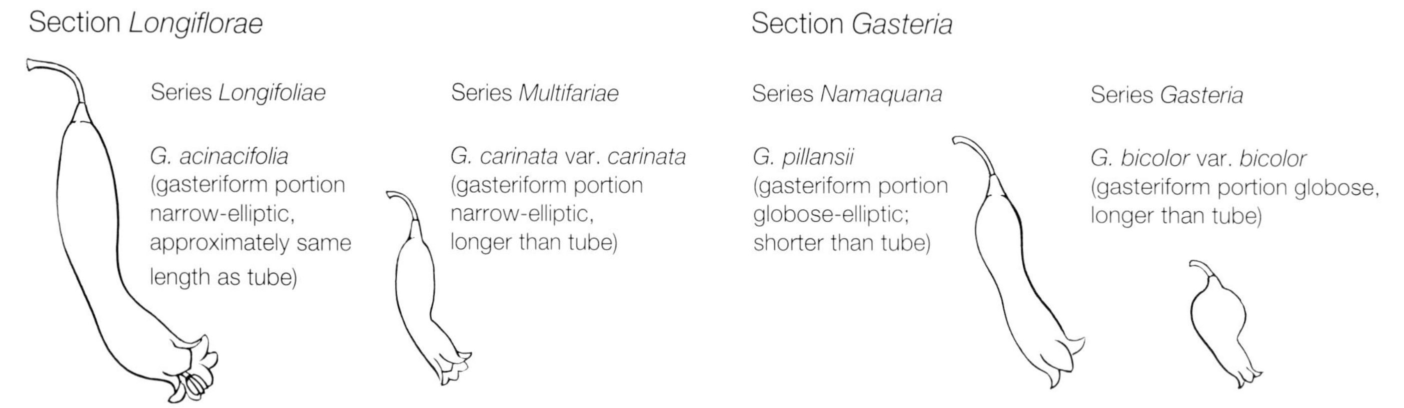

Perianth shapes of *Gasteria* flowers.

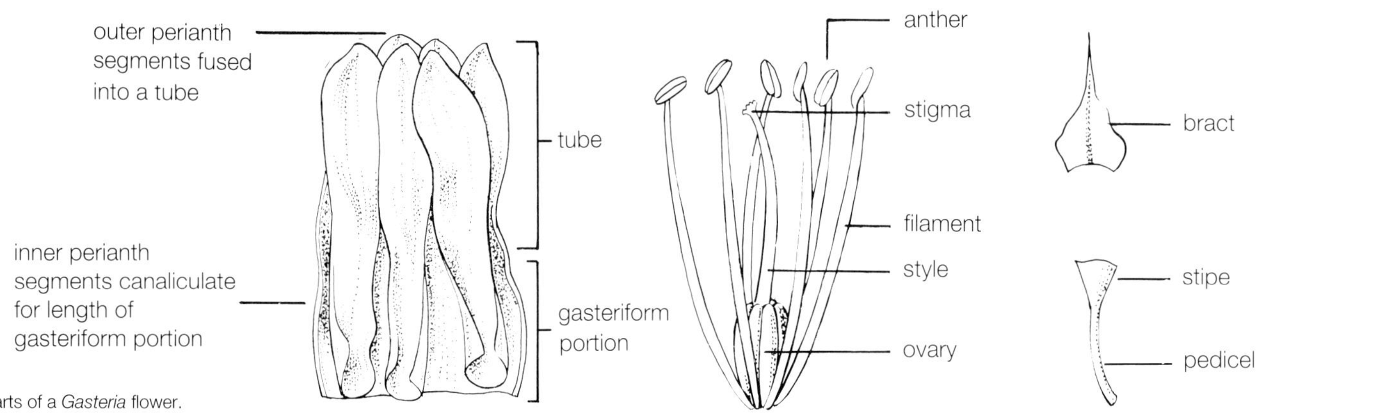

Parts of a *Gasteria* flower.

with acuminate apices. The midrib is somewhat keeled. The floral bracts are scarious, lanceolate and acuminate, and may be longer or shorter than the pedicels.

Perianth

The pendulous protandrous flowers are alternately and laxly arranged on pendulous recurved pedicels, usually secund when the racemes are spreading. However, the flowers of *Gasteria disticha*, *G. brachyphylla* and *G. bicolor* are often dense and, when the inflorescence is erect, are arranged equilaterally. The flowers of *G. acinacifolia* from the Port Elizabeth district are also densely arranged, but these are always secund. In the developing inflorescence, the pedicels and perianths are at first erect but soon become pendulous during flowering, after which the pedicels become arcuate, ascending when the fruit develops. The perianths, which are shortly stipitate at the base, are six-lobed. The segments are fused into a curved tube which is gasteriform at the base for up to two-thirds of the perianth length.

The length and diameter of the gasteriform portion in relation to the length and diameter of the tube is of importance in delimiting sections and series. The perianth is constricted below the globose or narrow elliptic gasteriform portion. The shape and size of the perianth in most species vary considerably. In *Gasteria pillansii* the flowers are often slightly clavate (above the gasteriform portion the tube dilates gradually to the same diameter or more as the gasteriform portion). The perianth in most cases is tricoloured; the gasteriform portion is dark to light pink, rarely white. The white tube has distinct green median striations on the lobes. *G. nitida* and *G. vlokii* have a dark reddish pink perianth and the apices of the former are yellowish. The inner segments are dorsally fused to the outer ones, and the membranous margins are free and canaliculate for the length of the gasteriform base and are adpressed to the ovary. The three channelled inner segments are filled with a clear nectar. The gasteriform portion could also, perhaps, serve as a functional protection for the ovary against phytophagous insects and bird damage. At the Kirstenbosch National Botanical Garden, slugs have been observed feeding on the stamens and segment apices, but the narrow tube stops them from gaining access to the ovary. Damaged flowers have not been observed in nature. The nectaries are situated directly below the ovary.

Stamens

The six filiform, slightly flattened filaments arise from just below the ovary and are nearly as long as the tube; occasionally one to three advance close to the mouth. In *Gasteria pillansii*, as well as rarely in *G. acinacifolia*, the filaments may be exserted by 4 mm. At the end of anthesis the filaments become contracted. The dorsifixed epipeltate anthers are oblong, dehiscing longitudinally.

Ovary

The oblong ovoid ovary is similar in all species except *Gasteria acinacifolia* and *G. rawlinsonii*, where it may possess three apical appendages. The style is uniform in all species. The trilobed stigma is minute at first, becoming capitate and lengthening to become shortly exserted or included at the end of anthesis. The stigma becomes moist when ripe, often with a characteristic clear droplet at the apex of the stigma.

Capsule and seeds

The three-angled capsules are loculicidal, oblong, woody and more or less uniform in all species except *Gasteria acinacifolia*, where they are markedly larger and truncate at the apex. The seeds are flattened, oblong, black and irregularly angled.

Gasteria seeds are variable, ranging from 2 to 8 mm in diameter. The surface characteristics of seeds have been used in many groups as a successful tool in plant classification (Bewley & Black 1986). Aloe seeds, on the other hand, have been investigated by Kamstra (1969, 1971), who looked at similarities which would allow grouping. However, he found that the differences were not sufficiently constant to be useful for classification purposes.

DESCRIPTIVE TERMINOLOGY

Perianth. The term 'gasteriform' has been used throughout to describe the spherical or inflated portion of the perianth base, at the suggestion of Dr John Rourke of the Compton Herbarium.

Leaf epidermis texture. The term 'tuberculate' refers to domed, hemispherical or convex, white or green, multicellular, regular or irregular outgrowths on the leaves of many *Gasteria* species. These tubercles are rarely secondarily tuberculate.

The term 'asperulous' refers to centric conical or subconical outgrowths, terminating in a spherical translucent papilla, of the outer walls of single cells. These outgrowths give a characteristic rough 'shark-skin' leaf texture found in several species.

Part of a *Gasteria* infructescence, showing the loculicidal capsules and the seeds.

Reproduction and seed dispersal

Gasteria species reproduce both by seeds, usually from cross-pollination, and vegetatively by stolons, proliferation from the base of the stem and broken leaf pieces. Seeds seem to play the most significant role in the dispersal of plants and contribute to gene flow, which is so important in an unpredictable and changing climate. The extreme genetic variability exhibited by most species may, in part, be attributed to successful reproduction by seed.

FLOWERING PERIOD

There appears to be little correlation between flowering time and the rainy season. Some taxa, as found in most South African aloes, flower in the dry season, and this seems to be an adaptation to enable the seeds to be dispersed in time for the following rainy season.

The flowering of some species that occur in all-year rainfall areas is erratic, with a peak in spring (*Gasteria rawlinsonii, G. carinata* and *G. bicolor*). *G. croucheri, G. batesiana* and *G. acinacifolia*, growing in a summer rainfall area, flower in summer (October to January), while *G. nitida, G. excelsa, G. vlokii* and *G. brachyphylla* var. *bayeri*, from an all-year rainfall area, flower in December and January. *G. pillansii*, which occurs in a winter rainfall area (Namaqualand), flowers from December to January and *G. pillansii* var. *ernesti-ruschii* from the northern Namaqualand coast in March and April. *G. disticha* from the winter rainfall area in the south-western Cape flowers in spring, whereas forms from Beaufort West flower in December.

POLLINATION

Self-incompatibility among Monocotyledonae appears to be a common phenomenon (Brewbaker & Gorrez 1967). All *Gasteria* species, with the exception of a few individual clones, appear to be obligate outbreeders, which may account partly for their genetic variability. Marshak (1934) found some clones of *G. disticha, G. carinata* and *G. bicolor* to be self-compatible. The author discovered that *G. vlokii* was also self-compatible. However, it was later found that this phenomenon was limited to one clone.

Gasteria flowers are visited by sunbirds (Nectariniidae) as well as insects, with the sturdy scape and curved, reddish pink, tubular perianth suggesting that the former play a more significant role. The conspicuous, erectly spreading, reddish pink racemes often protrude above the low karoo shrubs and are thus clearly visible and attractive to the birds. The flowers open acropetally, with three to five flowers per raceme usually open at the same time, representing various stages of development. This seems to favour cross-pollination as it encourages a bird to visit other plants (traplining). Flowering lasts about 30 to 45 days.

When feeding, a bird perches on the scape, usually sitting on top or, occasionally, hanging from below. Its curved beak is pushed neatly into the pendulous perianth which is then lifted into an erectly spreading position and the nectar is sucked up. This process is very rapid, only one to two seconds being spent on each flower. The gasteria plants in the shade house at Kirstenbosch are regularly visited by sunbirds during flowering, at about one- to two-hour intervals.

The bisexual flowers are protandrous, the anthers usually of irregular length, with one to three advancing in length. The dehiscing anthers lie at the bottom of the mouth with the pollen-shedding surface facing upwards, thus providing a maximum surface area. Pollen is rubbed off onto the dorsal side of the bird's beak while it is feeding and is transferred to the stigma of other plants, thereby ensuring cross-pollination. Nectar is exuded in copious quantities from the hypogynous nectaries, filling the three-channelled inner perianth segments (nectar channels). The apex of the style, which is also situated dorsally in the mouth of the perianth, may remain included or become slightly exserted when ripe, with the stigma incurved. Towards the end of anthesis the gasteriform portion of the perianth deflates somewhat and the clear, slightly viscous nectar may be guided to the mouth of the perianth where it is exuded as a characteristic droplet. The honey bee (*Apis mellifera*) and

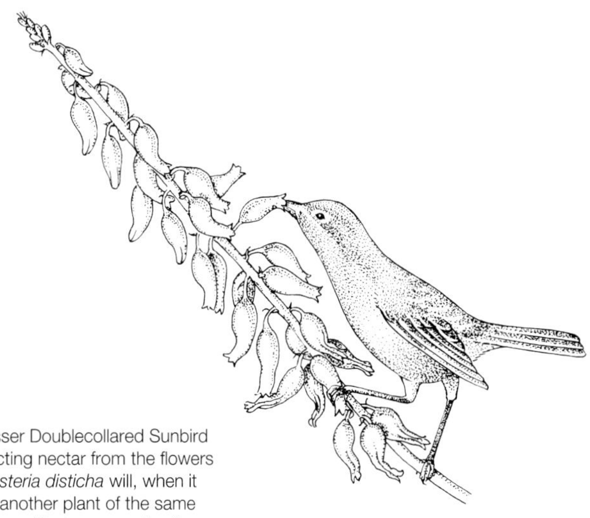

A Lesser Doublecollared Sunbird extracting nectar from the flowers of *Gasteria disticha* will, when it visits another plant of the same species, effect cross-pollination.

other bee species have been observed visiting the flowers. Each flower remains open for three to four days: from one to three anthers ripen during the first day, succeeded by the remaining anthers on the second day. The stigma ripens on the third day, when it becomes capitate and moist.

During 1988, breeding experiments on *Gasteria* species were initiated. The gasteria shade house at Kirstenbosch was netted to prevent possible cross-pollination. All species were self-pollinated at first but the flowers were soon aborted. Subsequently, all species were crossed with each other successively, using each species as the mother plant (pollen receiver). Capsule production and seed set were excellent (about 95 per cent seed set), and the few that failed were not thought to be due to incompatibility but possibly to transmitting pollen at the wrong time. A few individuals were then crossed with *Haworthia tenera*, *Aloe ferox*, *A. grandidentata*, *A. tenuior* var. *rubriflora* and *Bulbine latifolia*. The first- and last-named species proved to be unsuccessful (pollen donors and recipients), confirming Cutler & Brandham's finding (1977) that *Gasteria* and *Aloe* species hybridize readily, but those of *Gasteria* and *Haworthia* do not. Seed set in *Gasteria* hybridized with these *Aloe* species proved to be highly successful, also corroborating their close relationship. However, Rowley (1967, 1982) reported five intergeneric hybrids between *Gasteria* and other members of the Alooideae (*Aloe*, *Astroloba*, *Chortolirion*, *Haworthia* and *Lomatophyllum*). Finally, the flowering times of each species were recorded, along with the period involved from the time of pollination to when the capsule dehisced. It was found that seeds ripen within eight to ten weeks, on average, after pollination.

SEED DISPERSAL

The relatively tall, erectly spreading fruiting branch (infructescence), the erect capsule and winged seed are clear adaptations to wind dispersal. This method of dispersal falls into Van der Pijl's (1982) category of 'flyers'. The flat wings provide the means for gliding and those seeds with the largest wings have the widest distribution (*Gasteria acinacifolia*, *G. batesiana*, *G. pillansii*), while seeds with small wings have restricted ranges (*G. vlokii*, *G. rawlinsonii*, *G. pulchra*). The fruiting time of most species coincides with the rainy season or just before, whereas in species that occur in an all-year rainfall area, fruiting times are erratic, reflecting the unpredictability of the climate.

VEGETATIVE REPRODUCTION

Once a plant is established from seed, vegetative reproduction in most species usually occurs either by means of proliferation from the base or by subterranean stolons to form dense groups. There seems to be an inverse correlation between sexual and vegetative reproduction. *Gasteria* species that are very prolific from the base tend to have fewer inflorescences with fewer flowers, and not all individuals of the clone will flower, compared with species like *G. excelsa*, which is not prolific from the base but develops very large paniculate inflorescences carrying up to 2 700 flowers.

The tendency to prolific vegetative reproduction is probably also a means to survive predation by herbivores, as gasterias are frequently grazed by cattle, small antelope, rock rabbits (*Procavia capensis*), tortoises (*Geochelone pardalis*, *Chersina angulata*, *Homopus areolatus*) and porcupines (*Hystrix africaeaustralis*). The last-mentioned animals cause extensive damage in uprooting the plants. Parts remain in the soil and broken leaves proliferate to form new plants. This is true of all *Gasteria* species except *G. rawlinsonii*. The leaves of most species are brittle, especially those of *G. excelsa*; the broken pieces lie on the ground, take root and proliferate to form new plants. A population of *G. excelsa* was seen near Chalumna in the eastern Cape in an overgrazed river valley in which most plants had been extensively damaged by browsers and in which young plants regenerated from pieces of broken leaves. Vegetative propagules have also, though rarely, been observed on *G. acinacifolia* and *G. excelsa* inflorescences.

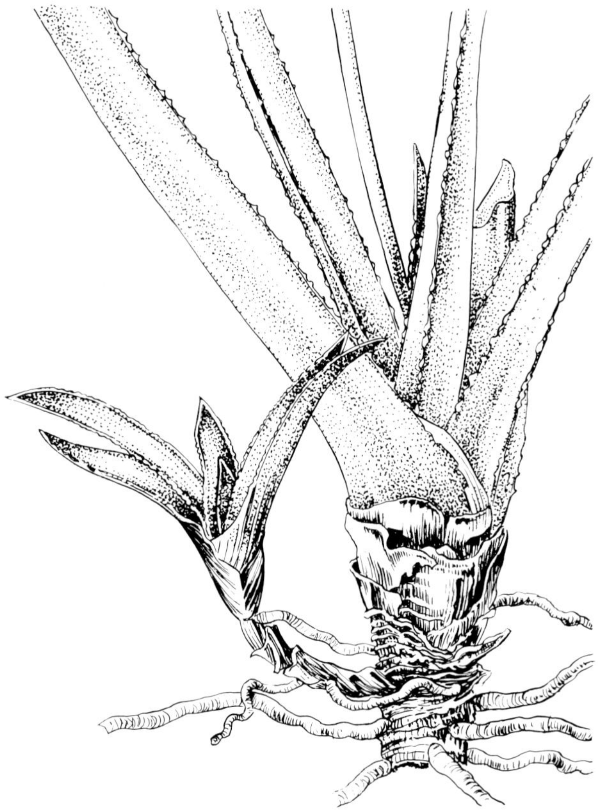

Vegetative reproduction from a stolon (*Gasteria pulchra*).

A rooted leaf of *G. excelsa* proliferating from the base.

In the Swartwaterspoort in the eastern Cape, a population of *Gasteria bicolor* var. *liliputana* was observed. As a result of heavy predation, believed to be by rock rabbits (dassies), very few undamaged adult plants could be found, yet the plants were genetically variable in leaf size and arrangement (phyllotaxy). Only one individual was found in flower. It seems that the plants may have a high mutation rate, giving rise to some of the aberrant forms so often found in a population.

In cultivation, damaged leaves may proliferate, forming clusters of juveniles on the damaged parts above the ground (*Gasteria batesiana*, *G. acinacifolia*).

Ecology

Two of the world's six major floristic regions are found in South Africa. The larger summer rainfall flora in the north forms part of the Palaeotropical Kingdom, and the flora of the predominantly winter rainfall region of the south and south-west forms the Cape Floristic Kingdom – the smallest floristic kingdom in the world. Where winter and summer rainfall overlap there is usually a mosaic, represented by species of both floral regions, and it is in one of these regions (the eastern Cape) that most *Gasteria* species can be found.

The flora of the winter rainfall region is the richer of the two in terms of species per unit area (Bond & Goldblatt 1984), and a very high proportion of it is endemic, with examples in the families Mesembryanthemaceae (the largest South African plant family), Proteaceae, Restionaceae, Ericaceae, Rutaceae and Fabaceae. Certain representatives of the flora are thought to be ancient, their origins possibly extending back to the Gondwanan supercontinent (Goldblatt 1978; Werger 1978). The summer rainfall flora is mainly of subtropical or tropical origin, with many species having a distribution range that extends northwards (Acocks 1953; White 1983). The active growing season coincides with the summer rainfall. Both regions can be divided into low and high rainfall regions.

CLIMATE

The South African arid climate is partly a result of the prevailing high pressure system over southern Africa, caused by the Hadley cell and intensified by the cold Benguela Current on the west coast. In the south the Ferrel westerly winds spiral eastwards causing the characteristic low-pressure systems that bring the winter rainfall to the southern parts of South Africa. Summer rainfall is caused by moisture-bearing south-east trade winds, which result in thunderstorms and other convectional rain in the inland and eastern parts of South Africa.

The climate of the eastern and south-eastern coastal plain is subtropical (eastern Cape and Natal), while the karroid portions of the southern and western Cape karoo regions have a warm temperate climate, usually with cool nights. The sparse vegetation of these karroid areas is dominated by dwarf shrubs of the families Mesembryanthemaceae, Asteraceae and Chenopodiaceae – comprising a high proportion of succulents – interspersed occasionally with taller shrubs of the families Anacardiaceae, Ebenaceae and Bignoniaceae. Towards the east the climate becomes more subtropical and the karroid vegetation is gradually replaced by arid subtropical thicket in the eastern Cape. The high altitude areas of the eastern parts of South Africa consist of a warm temperate grassland, usually devoid of trees, while in the sheltered southern and eastern mountain kloofs are small patches of afromontane forest.

Gasteria is a floristic component of the coastal arid, semi-arid and dry summer regions of South Africa below the climatically severe inland escarpment, predominantly influenced by winter rainfall. It also occurs to a lesser extent in the low-lying, hot dry river valleys (arid savanna) of Transkei, Natal, Swaziland and the eastern Transvaal, where precipitation occurs mainly in summer.

Based on the dominant plant life forms (the biome concept), the dry winter rainfall areas where *Gasteria* species are encountered can be divided into succulent karoo, renosterveld and fynbos. *Gasteria* is very well represented in the first two, but less so in the last. The dry summer rainfall areas can be divided into the Nama karoo, Great Karoo (Nama karoo) and arid savanna (valley bushveld). *Gasteria* features prominently in the last-mentioned area, but less so in the Great Karoo, and is absent entirely from the summer rainfall Namib regions. In the south-eastern Cape, where there is a mosaic of arid savanna and succulent karoo, there is a marked increase in the number of *Gasteria* species.

Gasterias are found from close to sea-level up to an altitude of approximately 1 700 m. Where frost occurs, it is light. Temperatures in summer are often very high, occasionally above 40 °C, but nights are usually cool in the southern and western parts. Rainfall of 25-800 mm per annum has been recorded in the habitat.

HABITAT

Gasterias grow on well-drained, humus-rich, sandy and sandy-loamy soils derived mainly from Palaeozoic quartzitic sandstone, sandstone and shales of the Cape Supergroup, as well as mudstones, shales and sandstones of the Karoo Sequence. To a lesser extent they also grow on shales, sandstones, and quartzitic sandstones of the Precambrian Malmesbury Group and Gariep Complex, as well as on Cambrian granites (Cape Granite Suite), Cretaceous conglomerates, Quaternary and Tertiary sands and alluvium.

The typical habitat consists of dry rocky hillsides inhabited by herbs with taller emergent shrubs under which gasterias are usually encountered. Plants are frequently found in rock fissures or in the shade of rocks, with an occasional plant in the open. *Gasteria pillansii*, *G. disticha* and *G. brachyphylla* are usually associated with the succulent karoo (White's [1971] Karoo-Namib domain) where succulents of the family Mesembryanthemaceae predominate. The vegetation is open, with dwarf succulent shrubs and occasional taller emergent shrubs. The species *G. disticha* and *G. brachyphylla* are also found in renosterveld with its preponderance of *Elytropappus rhinocerotis* (renosterbos, a shrub of 600-1 000 mm in height). *G. pillansii* is the only species confined solely to a winter rainfall region, while the two previously mentioned species occur in a region that receives rain throughout the year. *G. carinata* occurs in renosterveld (White's [1971] Capensis domain) which receives rainfall throughout the year, but is also occasionally encountered in southern outliers of valley bushveld (arid savanna), a dense, thorny vegetation type of the eastern Cape. *G. vlokii* is somewhat unique in habitat as it occurs in shallow east- and north-facing rocky pockets and fissures of steep quartzitic sandstone of the Table Mountain Group, up to an altitude of 1 500 m. There the soils are acidic and poor in minerals, and the vegetation consists of mesic fynbos.

DISTRIBUTION AND SPECIES DIVERSITY

The present distribution of gasterias is confined to the coastal parts of South Africa, below the escarpment and usually not more than 200 km from the coast. *Gasteria* species are found from just north of the Orange River (southern Namib) in the north-west to Zululand in the north-east, and along the dry river valleys of Swaziland and the eastern Transvaal. *G. pillansii* is confined to Namaqualand, from Clanwilliam in the south to the southern Namib in the north. *G. batesiana* has a disjunct distribution – the northernmost of the genus – occurring in isolated river valleys in the north-eastern Transvaal, Swaziland and Zululand in northern Natal. *G. acinacifolia* (from the eastern Cape) and *G. croucheri* (from Natal), two closely related vicariant species, are closely allied to *G. batesiana* and together have the most widespread distribution, ranging from southern Zululand in the north (Natal) to Knysna in the south (Cape).

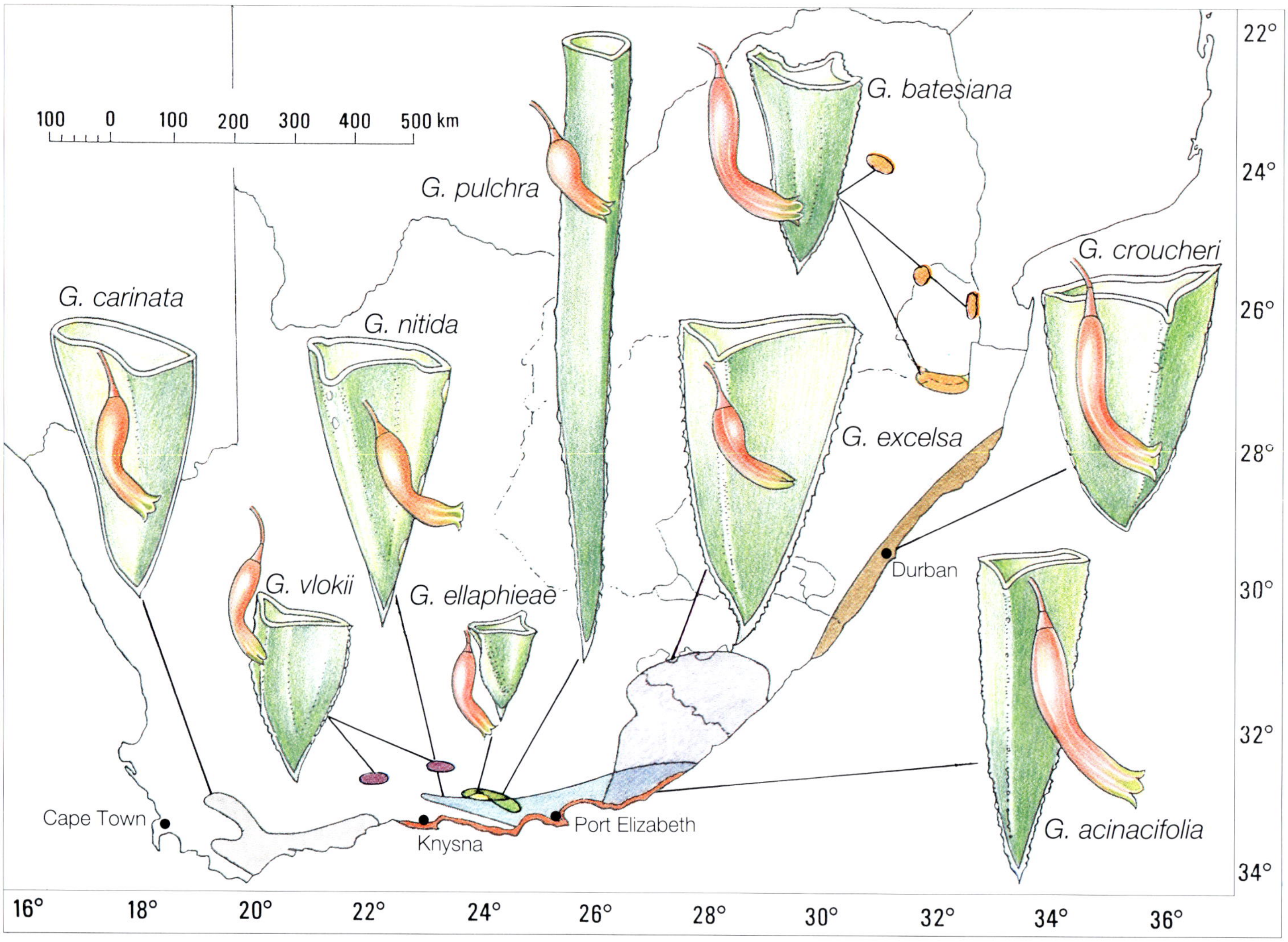

Map showing the distribution of *Gasteria* species in the section *Longiflorae*.

The remaining 12 species have a fairly restricted distribution, confined to the southern and eastern Cape. In the southern Cape four species occur, of which three are clearly separated by the quartzitic sandstone mountains of the Cape Supergroup. *Gasteria carinata* is confined to the coastal flats south of the Langeberg and Outeniqua ranges. *G. brachyphylla* occurs between the Langeberg and the Outeniqua ranges in the south, and the Swartberg range in the north. *G. disticha* is found north of the Swartberg range and south of the escarpment mountains to the north, as well as in the Breede River valley and at the western end of the Little Karoo. *G. vlokii* is confined to the top of the Swartberg between the Little and Great Karoo. *G. bicolor* is widespread in the south-eastern Cape, while *G. baylissiana* is endemic to the Suurberg north of Port Elizabeth. *G. rawlinsonii* is a narrow endemic of the Baviaanskloof. *G. ellaphieae* and *G. glomerata* are endemic to the Kouga Dam region.

The remaining three related species also occur in the south-eastern Cape. *Gasteria nitida* is widespread from Uniondale in the west to Bathurst in the east, while *G. pulchra* is endemic to the lower Gamtoos Valley. *G. excelsa* occurs from the Grahamstown district in the south-eastern Cape to southern Transkei.

The south-eastern Cape's diversity in topography, geology, climate and vegetation explains, to a degree, the species diversity of *Gasteria* in the region (Cowling 1982; Gibbs Russell & Robinson 1981). Rainfall occurs in winter and summer in this semi-arid area; there is, however, a tendency to summer dryness in the south-west and winter dryness in the east. Vegetation types that merge here include fynbos, succulent karoo, karoo, arid savanna (valley bushveld), lowland forest, afromontane forest and grassland (White's [1971] Capensis, Karoo-Namib, Sudano-Zambezian and Usambara-Zululand domains). Coastal hills and an inland plain, with wide river valleys divided by mountain ranges of the Cape Supergroup, make up the terrain, which consists mainly of Cretaceous conglomerates and Tertiary alluvium, and in the east of shales of the Karoo Sequence. The vegetation (valley bushveld, comprising a dense thorny thicket that is rich in succulent species) seems to be a floral mosaic representing the arid savanna and succulent karoo. It is the habitat of *G. bicolor*, *G. croucheri*, *G. pulchra*, *G. baylissiana*, *G. excelsa*, *G. glomerata*, *G. ellaphieae* and *G. rawlinsonii*.

The mountains, which consist of quartzitic sandstone of the Witteberg or Table Mountain Group (Cape Supergroup), are low in mineral and nutrient compounds and usually carry grassveld or fynbos, and to a lesser extent, valley bushveld and renosterveld. This more open, grassy habitat is the home of *G. nitida*, which usually occurs on well-drained slopes in full sun, or sometimes in light shade. The grassveld and fynbos are subject to fires but this species will regenerate from damaged parts. *G. acinacifolia* is found on the southern and eastern coastal dunes on alkaline littoral soils, but also as a cliff-dweller at Knysna.

When the present distribution of all the species is analysed, one sees the following: of the 16 species, the four with the most primitive features have the most widespread and northern distribution (*Gasteria batesiana*, *G. acinacifolia*, *G. croucheri* and *G. pillansii*). All 16 species have distributions which are clearly allopatric, and when sympatric the habitats are not shared. Species richness increases dramatically in the eastern Cape, a conflict zone between summer and winter rainfall, where nine of the 16 species can be found. The highest concentration occurs in the Steytlerville degree square, where seven species occur. In

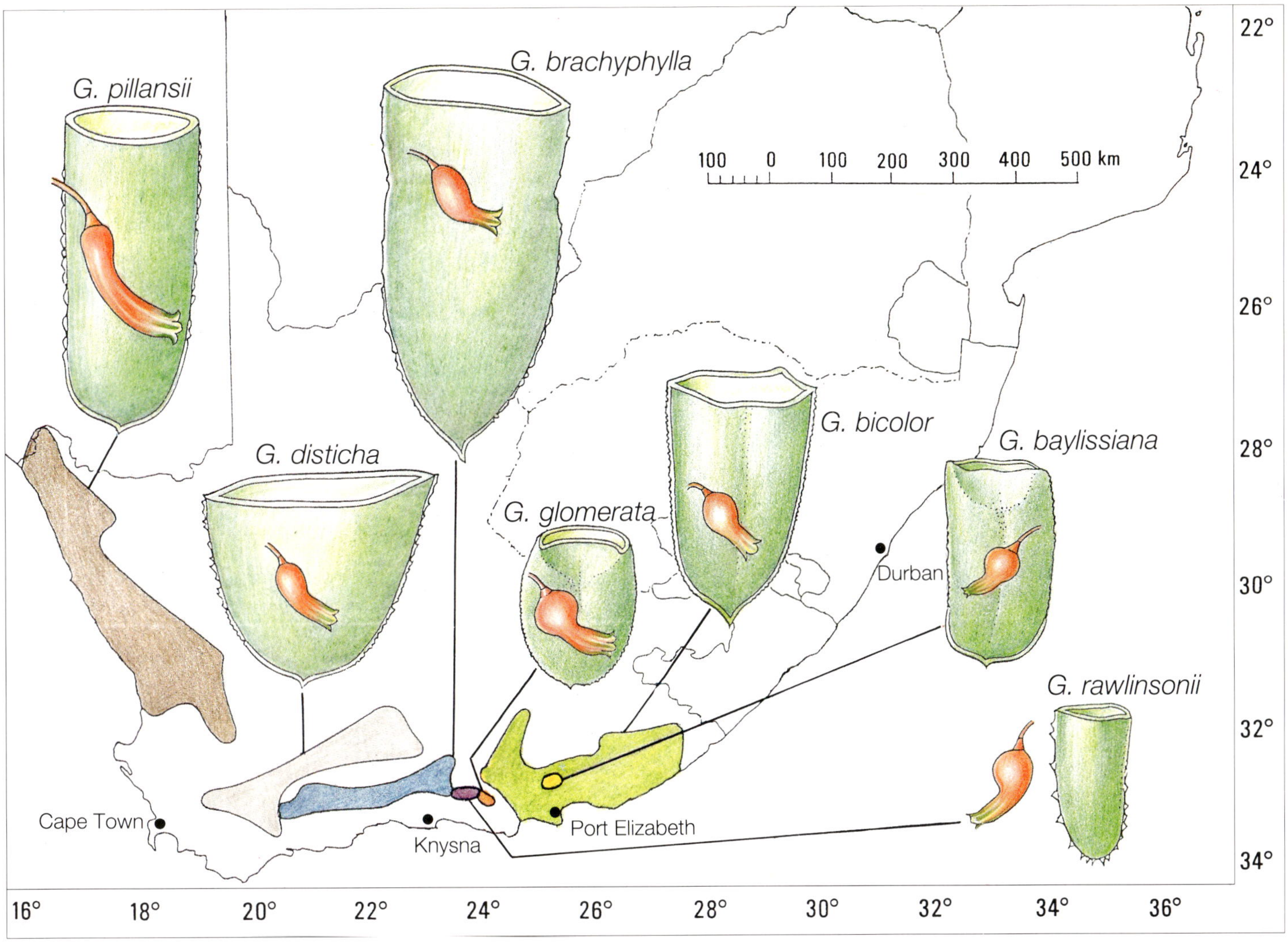

Map showing the distribution of *Gasteria* species in the section *Gasteria*.

G. bicolor, *G. brachyphylla*, *G. disticha*, *G. pillansii* and *G. carinata* there is a marked increase in genotypic variation associated with terrain diversity (especially with rock formations of the Cape Supergroup). A number of other typical Cape taxa show similar distribution and speciation trends to that of *Gasteria*, having species with primitive features and widespread distribution to the north, and with a dramatic increase in species richness in the southern Cape. These taxa include *Protea*, *Erica*, *Bulbine*, *Crassula*, *Cotyledon* and *Delosperma* (Weimarck 1941; Goldblatt 1978; Linder 1983).

SURVIVAL IN AN ARID ENVIRONMENT

Water is stored in the succulent leaves, which become deeply channelled during drought but recover rapidly after rain. The very shallow, spreading root system enables plants to utilize light showers effectively. The roots of all *Gasteria* species are succulent but they are linear obclavate in *G. nitida*, *G. ellaphieae* and *G. vlokii*, thus enabling plants to store ample water. *Gasteria* plants are found usually on southern and eastern aspects of koppies, ridges and mountain slopes, where they are protected from the hot afternoon sun.

Most species occur beneath xerophytic herbs and shrubs, or in the shadow of rocks, thus ensuring sufficient shade and also a constant organic food supply. The shallow spreading root system also enables plants to feed from the rich upper soil layer.

DEFENCE MECHANISMS

Teleologically considered, predators are effectively avoided – or the effects of their predation turned to advantage – in a number of ways.

The leaf margins of most *Gasteria* species are armed with white cartilaginous tubercles which can be serrulate, often continuous and very sharp. The cartilaginous texture may inhibit predators, while the dense tuberculate leaf surface of *G. carinata* var. *verrucosa* and *G. baylissiana*, and the asperulous epidermis of *G. disticha* and *G. rawlinsonii* could serve as a protection against phytophagous insects (Van Jaarsveld 1987b).

Most *Gasteria* species grow within the 'arms' of other, thorny plants where they find not only protection from grazing herbivores, but also shade and an organic food supply. *G. brachyphylla* var. *bayeri* is sheltered by *Euphorbia mammillaris*.

Gasteria ellaphieae blends in with the natural quartzitic sandstone of the Kouga Dam area, eastern Cape.

The serrulate margin of *Gasteria excelsa* is razor-sharp, providing a defence against grazing animals.

During the course of field study, no plants with any signs of insect predation were observed.

Most *Gasteria* species have spotted leaves which serve as camouflage in the shrubby plant cover. The leaves of *Gasteria nitida* var. *armstrongii* resemble the stony background of the plant's habitat.

Gasteria rawlinsonii and *G. glomerata* are cliff-dwelling species of the sheer narrow ravines of the Baviaanskloof and Kouga mountains – out of reach of most larger herbivores. These species may be only very faintly spotted, corroborating the hypothesis that the spots evolved as a camouflage strategy in response to herbivory (Van Jaarsveld 1987b). *G. croucheri* is also found on sheer, inaccessible rock faces of the dry river valleys of southern and central Natal.

In habitat, most *Gasteria* species occur in the 'protective arms' of thorny shrubs and herbs. *Carissa haematocarpa* and *Azima tetracantha* are dense, fiercely armed, rigid shrubs with which gasterias are frequently associated.

Although some populations are heavily preyed upon, the brittle leaf pieces that drop to the ground root and regenerate rapidly, thus ensuring the survival of the species despite the loss of the parent plant. *Gasteria rawlinsonii* is the only species that does not proliferate from broken leaf pieces, corroborating the hypothesis that vegetative reproduction evolved in response to herbivory. There is a great difference in the growth rate between juveniles that are grown from seed and those grown from leaves. Plants grown from seed may reach maturity in about three to five years, whereas a juvenile grown from a single mature leaf may flower within two years. The rapid attainment of maturity may be explained by the parent leaf rooting rapidly, acting as a food resource and conveying nourishment to its progeny.

The extreme genetic variability of this neo-endemic genus ensures a rapid adaptation to change in environmental conditions as well as the pioneering of new habitats. Their phenotypic plasticity ensures that individuals will adjust to short-term environmental changes – for instance, to the degree of change in light. *Gasteria nitida* retains a state of juvenility when there are frequent fires in its habitat.

Conservation

Fortunately no known *Gasteria* species has become extinct, and most species are still relatively common in their natural environment. The ease with which these plants propagate naturally from leaf fragments or offsets has contributed to their survival to date. However, the destruction of *Gasteria* habitat as a result of agricultural practices and the indiscriminate collecting of plants by herbalists and succulent collectors have necessitated the present conservation legislation.

All *Gasteria* species are protected in habitat in South Africa, and plants may not be removed without a permit from the appropriate regional authority. Traditional healers and herbalists should be encouraged to cultivate and maintain their own stock or to obtain plants from legal authorities instead of collecting in the veld. Even if they do collect in the veld, they should take leaf cuttings or offsets rather than the whole plant, so that genetic variability may be maintained as far as possible. Constructive conservation education is needed in order to motivate and stimulate the public to conserve and cherish plants and their habitats.

THE CURRENT CONSERVATION STATUS OF *GASTERIA*

Two *Gasteria* taxa, *G. nitida* var. *armstrongii* and *G. bicolor* var. *liliputana*, have been listed in *Threatened plants of southern Africa* (Hall *et al.* 1980), and their status has been given as 'uncertain'. Both species occur in the eastern Cape.

Threatened habitats

Most *Gasteria* species occur in the south-eastern Cape, where there are a number of reserves and natural areas in which they are protected. However, in many instances the reserves are small and offer a refuge to no more than a fragment of the species population. In order to conserve the greatest diversity within the species' gene pool, the plants must also be protected outside reserves.

Some species, notably *Gasteria excelsa*, *G. carinata* var. *carinata*, *G. carinata* var. *retusa*, *G. pulchra*, *G. nitida* var. *nitida*, *G. pillansii* var. *ernesti-ruschii* and *G. baylissiana*, do not occur in reserves at all. Of these, *G. pillansii* var. *ernesti-ruschii* and *G. carinata* var. *retusa* have a fairly restricted distribution, but are fortunately reasonably common and are not threatened at present.

In the savanna biome of the south-eastern Cape the arid region, which includes valley bushveld, noorsveld and dune thicket, has been extensively transformed into cultivated fields or has suffered severely from overgrazing. *Gasteria bicolor*, *G. pulchra*, *G. acinacifolia*, *G. excelsa*, *G. rawlinsonii*, *G. baylissiana*, *G. ellaphieae* and *G. glomerata* all occur in this habitat.

Although large portions of the natural habitat of *Gasteria bicolor* in this part of the savanna biome have been destroyed by agriculture and urban expansion, this species is still common and widespread. *G. excelsa* is less common, although also widespread, and needs protection as its populations are threatened by grazing. *G. acinacifolia* grows on coastal dunes and is relatively safe.

Most other savanna species of the eastern Cape occur in specialized habitats and are not likely to be eradicated by ploughing. These include *G. pulchra*, which is restricted to conglomerate soils in the hilly regions of the Hankey and Patensie districts (an area that hosts numerous other succulent species and should be granted formal protection status); *G. glomerata* and *G. ellaphieae*, which are restricted to the Kouga Dam area in the protected Baviaanskloof region; *G. rawlinsonii*, which grows mainly in a protected zone, but where plants do occur on private farms they are inaccessible; and *G. baylissiana*, a critically rare species which so far is known only from the top of the Suurberg (Van Jaarsveld 1990, 1991a). This last species has been increased considerably through a propagation programme at Kirstenbosch National Botanical Garden and has recently been replanted in its habitat with funds provided by the Cactus and Succulent Society of Great Britain. Plants have also been obtained from the original stock given by Bayliss to various botanical gardens where it has been sexually increased.

In the grassland biome *Gasteria nitida* var. *nitida*, the only species that occurs there, is relatively widespread in the eastern Cape. Nevertheless, a protected area where it can grow undisturbed is needed to safeguard its future. This species is also found in the renosterveld of the eastern Cape.

Two taxa which are confined to renosterveld are *Gasteria carinata* var. *carinata* and var. *verrucosa*. Forming part of the fynbos biome, the renosterveld is one of the most threatened vegetation types in South Africa, as much of it has already been transformed into wheat fields. Fortunately the two *G. carinata* varieties, together with other very rare taxa, are restricted to shale or calcrete outcrops that form 'islands' in the cultivated land. More sites are needed in which *G. carinata* in particular can be protected, as it is one of the most variable species, with many interesting and attractive forms. Another species of the fynbos biome is *G. vlokii*, which is largely confined to drier fynbos at the top of the Swartberg where its habitat is protected. However, recently this species has also been collected from other parts of the Great Swartberg range and from the Witteberg, near Willowmore. It is probably more widespread than was previously thought.

Gasteria pillansii, *G. disticha*, *G. brachyphylla* and *G. carinata* var. *retusa* occur in the succulent karoo and Nama karoo biomes and are still relatively common, the first three still being found in nature reserves. The only threat at present is overgrazing in some areas, which could lead to a decline in population sizes.

FUTURE CONSERVATION ACTION

As more and more demands on land are made for agriculture and urbanization, pressure on natural *Gasteria* populations will inevitably increase. At present the conservation status of most species appears to be adequate, but there is clearly a need for a strategy that will protect indefinitely the genus as a whole. Such a strategy should ensure that major *Gasteria* habitats are protected in their entirety, and that minimum requirements regarding the size of such habitat areas should be adhered to. Core conservation areas need to be identified by means of research into centres of *Gasteria* endemism and species concentrations, where possible linking these with other plant species which have a similar habitat specificity. An urgent priority is that sites facing imminent destruction for agricultural or urbanization purposes need to be identified and appropriate action taken, such as the preservation of the habitat in selected natural remnants.

Subtropical thicket in the eastern Cape, the habitat of most *Gasteria* species, is giving way to urbanization, agriculture and alien plants, in this case *Opuntia ficus-indica*.

Species that require immediate attention are *G. excelsa*, *G. pulchra*, *G. nitida* var. *nitida*, and *G. baylissiana* in the eastern Cape; the three varieties of *G. carinata* in the southern Cape; and *G. pillansii* var. *ernesti-ruschii* in the north-western Cape.

Much success has been enjoyed by *ex situ* conservation efforts, particularly at the Kirstenbosch and Karoo national botanical gardens, where all the species are under cultivation. *Gasteria* plants and seed are made available to traditional herbalists at Kirstenbosch and at Durban municipality's Silverglen Nursery.

Finally, no conservation strategy can be entirely successful without effective education of the general public, developers and farmers. Education should focus jointly on the need to protect the gasterias' unique and at times threatened habitat, and to popularize the species through horticultural initiatives. Annual plant sales at Kirstenbosch and the Karoo botanical gardens ensure that many *Gasteria* plants reach the homes of enthusiasts.

Cultivation

An age-old culture, plant cultivation arose first from Man's need for food, once he had abandoned his nomadic existence; later, with the expansion of his knowledge, some plants were grown specifically for their curative properties, but it is probably only in the last 4 000 years or so that Man's love of plants has led to their cultivation for ornamental purposes – and a mere 300 years in the case of gasterias. As a skill, cultivation is developed largely through experience, and gasterias, as some of the easiest plants to grow, are an ideal starting point for the novice. No 'green fingers' are required; one only has to follow the simple guidelines set out below and gasterias will grow very successfully.

Many plants (including certain *Gasteria* species) have become rare due to habitat destruction. Where this has happened, propagation has played, and continues to play, an important role in increasing the numbers of these plants or even reintroducing species to their natural environment. At least four gasterias have been considerably reduced in number due to farming activities and urban encroachment. But for the most part, fortunately, gasterias grow on rocky outcrops or vertical cliff faces, where agricultural activities pose no threat.

Plants can be reproduced either from seed or vegetatively from cuttings. Both are important avenues of plant breeding. Grown from seed, no two plants are identical, but grown from cuttings, all plants are replicas of the mother plant. Both methods of propagation are adopted in gasteria cultivation and conservation. Not all plant species, however, are easily cultivated; some are uniquely adapted to a specific set of conditions which are difficult to simulate, and in these instances a knowledge of and insight into a plant's habitat and natural distribution are essential. But there are other plants, such as gasterias, which adapt easily to a variety of habitats and soil types. One of the reasons for the ease with which gasterias can be propagated stems from the varied conditions of their southern and eastern Cape habitat.

Gasterias have been popular as pot plants since the late seventeenth century when they were first introduced into Europe. They are the easiest succulents to cultivate and thrive on windowsills and stoeps. They propagate readily from leaf cuttings or division, and even when a leaf drops from the mother plant it roots rapidly, forming small plants at the base of the broken leaf. Juvenile plants are very different from the adults

Gasteria excelsa in cultivation in a rockery at Kirstenbosch. The plants need a well-drained site and thrive when given an annual dressing of compost.

Gasteria bicolor var. *bicolor* is one of several species that will grow in a vertical stone wall.

and go through fascinating growth stages, starting with opposite leaves (some rough in texture) and later changing to a rosette and a smooth leaf surface.

Seeds and plants were sent to Europe by Governor Simon van der Stel after probably first having been cultivated in the Company's garden at the Cape. The plants, being hardy succulents, easily survived the long sea journeys, but they could not endure the severity of the European winters, and indoor cultivation became essential. They were found to thrive in conditions of diminished light intensity and low temperatures, and in a wide range of soils. These advantages, together with their attractive, mottled, strap-like succulent leaves and showy tricoloured flowers, not to mention their easy propagation and virtual freedom from disease, ensured their popularity.

OUTDOOR CULTIVATION

Gasterias can be successfully cultivated in a semi-arid to arid, a warm-temperate or a subtropical climate. They thrive in well-drained, sandy soil to which plenty of compost or leaf mould has been added. If the soil is not well drained, sand or grit should be added to improve its porosity. They can be grown in open, raised rockeries or as part of miniature rockeries, or they can be grown in containers on open patios, verandas and elsewhere. They grow successfully even at altitudes of 2 000 m, in warm, protected, north-facing positions (in the southern hemisphere), including on well-drained slopes. Most species prefer light shade and should be protected from direct sun and severe frost.

Certain species of *Gasteria* will, to some extent, tolerate direct sun. (The author has observed *G. carinata* and *G. bicolor* in full sun in gardens on the Witwatersrand.) However, never move to a sunny position plants that have been grown in the shade, unless they have first been hardened off. This is done by gradually reducing their time in the shade and increasing their exposure to the sun – a process that should take a few weeks. In high rainfall areas, weeds can be a problem when growing gasterias outside. An inorganic pebble mulch around plants will suppress weed growth and improve the appearance of a rockery. Various gravel sizes, textures and colours are also available. Once the rockery has stabilized, weeds will become less of a problem since they are opportunists that need disturbed soil and a high light intensity to germinate.

Gasterias transplant readily, but be sure to select a position that will be shaded by a rock, or wait until the nurse plant has grown large enough to provide the necessary amount of shade. Alternatively, a shade net can initially be stretched over the plants. Like aloes and haworthias, gasterias have an adventitious root system which should be pruned back to the stem before planting. They also need to be well firmed down in the soil. Most species have a shallow mat-like root growth which responds well to applications of fertilizer throughout the growing season. Organic material in the form of compost or leaf mould and ample bonemeal should be applied in spring to further enrich the soil. Most gasterias need water all year round, but it should be applied sparingly. In high rainfall regions it is not necessary to water. *Gasteria pillansii* comes from a winter rainfall region and therefore should be kept dry in summer.

Gasterias are ideal for growing in crevices of vertical sandstone walls, but here plants should preferably be planted when small. Suitable species for vertical walls include: *Gasteria rawlinsonii*, *G. batesiana*, *G. croucheri*, *G. excelsa*, *G. baylissiana*, *G. bicolor* and *G. glomerata*.

Planting a rockery

Natural 'rock gardens' are usually confined to raised or mountainous areas or river valleys. Therefore, when constructing a rock garden the key is to work in harmony with nature. Obviously, a natural rocky slope or outcrop is ideal but in their absence a rocky hillock can be created –

preferably to blend in with the rest of your garden - using local rock. The top soil with which you finish a rockery should be 150-300 mm deep, and of course be well drained and of a sandy loam. A well-balanced rockery consists of various types of drought-resistant shrubs, herbaceous perennials, bulbs, annuals and a number of different succulents. Avoid aggressive plants that will overgrow the rock garden, or be sure to prune back when necessary. Various shrubby succulents and xerophytes can be planted among gasterias to provide the light shade necessary for successful cultivation. For this purpose include the Karoo num-num (*Carissa haematocarpa*), granaatbos (*Rhigozum obovatum*), kerkei (*Crassula ovata*), *C. arborescens*, spekboom (*Portulacaria afra*), *Felicia filifolia*, renosterbos (*Elytropappus rhinocerotis*), and *Diospyros austro-africana*. Smaller shrubs that provide shade include *Crassula tetragona* subsp. *tetragona*, *C. sarcocaulis*, *Lampranthus* spp., *Ruschia* spp. and *Felicia heterophylla*.

INDOOR CULTIVATION

As gardeners of the late seventeenth and the eighteenth centuries discovered, gasterias can be grown very successfully indoors. Nowadays there are several cultivars available in most species and they are usually of varying sizes. It is therefore possible to build up a rewarding collection of the plants, even in limited space. Moreover, the size of the individual flowers remains the same, irrespective of the size of the vegetative parts, so that the smaller the plant, the more striking are the flowers.

Although gasterias will grow in almost any soil, the following simple mixture will provide good nourishment for any of the species: two parts sand, one part leaf mould or compost and one part loam (good garden soil). Add 250 ml bonemeal or superphosphate to one wheelbarrow of soil and mix the ingredients thoroughly. Gasterias also thrive in a mixture of half peat (or fern fibre) and half perlite, and they tend to grow more quickly in this medium - doubtless due to its excellent aeration - but a liquid fertilizer such as Seagrow should occasionally be added when watering. (Remember, a well-fed plant is more resistant to diseases than one that is undernourished.) The ease with which gasterias grow in this medium is an indication of its similarity to the plant's natural habitat where the roots occupy the upper soil layer, rich in leaf litter. Another advantage of this medium is that the peat inhibits the spread of fungal diseases.

Gasterias adapt to whatever light is available. With less light, the plants become greener and larger, while with more light the opposite result is produced. The plants will adapt better to indoor conditions if placed close to windows or on windowsills, but be careful of a hot, north-west window without the necessary shade. A greenhouse - no matter how small - is an ideal growing spot, particularly if the amount of shade can be altered according to the plants' needs.

Take care when moving plants from one position to another, since gasterias are sensitive to sunburn, especially when they have been grown in a shady corner and are suddenly moved into a sunny position. Even half an hour in full sun can badly damage a plant which has not been hardened off. The plants, and seedlings, must be weaned gradually from the shade.

Water the plants sparingly throughout the year, no more than once a week in summer and even less frequently in winter. A liquid fertilizer such as Seagrow can be added to the water every three months. If the leaves become deeply channelled and thin, it is normally a sign that the plant needs water; on the other hand, too much water can cause rot. *Gasteria pillansii* should be kept completely dry in summer.

Gasterias in pots and containers

Almost any type of pot is suitable for growing gasterias, as long as it provides sufficient drainage. In very wet climates, clay pots provide better aeration and drainage for plants growing out of doors. However, generally gasterias are not fussy. When starting a *Gasteria* collection, pots of the same size and shape produce a neater effect than an assortment, and rectangular plastic pots waste less space and look particularly trim. Many of the smaller species are adapted to grow in areas of very limited space and can be grown in very small (8-cm) containers, making them ideal gifts. The bottoms of pots should be lined with a layer of newspaper to prevent any soil from running out and to ensure good drainage. Alternatively, coarse gravel, small pebbles or pot shards can be used.

Before planting, prune back the old roots with sharp secateurs or a knife, and dust the freshly cut parts with sulphur to prevent soil-borne diseases from contaminating them. Next, fill each pot with potting soil, firming it down to within 10 mm from the top of the pot to provide space for watering. The gasteria can now be planted with only the basal portion submerged in the soil. Finish off with a layer of gravel to suppress weed growth.

Dwarf *Gasteria* species suitable for small containers (8 x 8 cm) include: *G. bicolor* var. *liliputana*, *G. pillansii* var. *ernesti-ruschii*, *G. brachyphylla* var. *bayeri*, *G. ellaphieae*, *G. glomerata*, *G. baylissiana* and *G. nitida* var. *armstrongii*. Some *Gasteria* species suitable for small to medium containers (9 x 9 cm) are: *G. carinata* var. *retusa*, *G. carinata* var. *verrucosa*, *G. vlokii*, *G. batesiana*, *G. rawlinsonii*, and small forms of *G. bicolor* var. *bicolor*, *G. pillansii* var. *pillansii* and *G. nitida* var. *nitida*.

Gasteria carinata var. *carinata*, *G. nitida* var. *nitida*, *G. pulchra*, *G. brachyphylla* var. *brachyphylla*, *G. disticha*, *G. bicolor* var. *bicolor* and *G. pillansii* var. *pillansii* are suitable for medium-sized containers (12,5 x 12,5 cm to 17,5 x 17,5 cm). *G. croucheri*, *G. bicolor* var. *bicolor*, *G. excelsa*, *G. pulchra* and *G. acinacifolia* can be grown in larger containers (from 23 x 23 cm).

Remember that most species are adapted to growing in limited space and are cluster-forming, but provided the plants are healthy and given sufficient food, division should not be necessary for three to four years.

Repotting

Repot plants every three to four years, and trim the roots back. If the plant has developed into a dense cluster, it can be divided with a sharp knife. Dust the damaged parts with sulphur or treat with a fungicide and allow them to dry out for about a week before replanting. Plant in a fresh soil mixture and discard the old soil unless it has been renewed with compost or leaf mould.

Miniature gardens

The smaller *Gasteria* species are ideally suited to a miniature garden in a container. Select whichever type of container you prefer, but ensure that there will be good drainage for the plants. Empty perlemoen (abalone) shells make very attractive containers and are sufficiently porous. Either plant the garden with a variety of small gasterias or combine a mixture of miniature succulents, using the recommended soil mixture. Firm the soil down to within 10 mm from the top of the

Rocks with fissures and hollows make excellent natural containers for small *Gasteria* species. *G. bicolor var. liliputana* and *G. baylissiana* grow well in a sandstone rock.

Gasteria species can be grown with other small succulents in a container to create an attractive miniature garden. In this example *G. bicolor* var. *liliputana* (left front) shares a garden with *Senecio rowleyanus, Crassula orbicularis, Delosperma elsieae, Faucaria felina* and *Haworthia translucens* var. *tenera.* The tall species in the centre is *Dietes grandiflora.*

container. Stones can be placed effectively among the plants to create a miniature rockery, if you so wish.

Suitable candidates for several plantings in one container are *Gasteria bicolor* var. *liliputana, G. brachyphylla* var. *bayeri, G. baylissiana, G. carinata* var. *retusa* and var. *verrucosa, G. glomerata, G. ellaphieae, G. pillansii* var. *ernesti-ruschii.* Other small succulent species suited to growing with *Gasteria* species include *Crassula pellucida* subsp. *marginalis, C. rupestris* subsp. *rupestris* 'Kouga', *C. orbicularis, C. lanuginosa* var.

To take a leaf cutting from a gasteria, hold the container or the plant itself firmly with one hand and twist the leaf sideways until it breaks away. A cutting will also root if it is cut from the parent plant with a knife.

pachystemon, C. muscosa, C. expansa subsp. *fragilis, Senecio radicans, S. rowleyanus, S. articulatus, Haworthia tessellata, H. translucens* subsp. *tenera, H. cooperi* and *Huernia echidnopsioides.*

PROPAGATION

One of the advantages of growing gasterias is that the plants reproduce readily and are easily propagated by means of seeds, leaf cuttings, offsets or division.

Tools and equipment

A small greenhouse or cold frame is the ideal place for propagation, but it is by no means essential. All species of *Gasteria* can as easily be propagated on a patio or on a warm windowsill. You will need a clean work bench, old newspaper, a trowel, dibbers (planting sticks), a watering can with a fine rose, a hand sprayer, a sharp knife, secateurs, planting trays, a pressure board (for firming down the soil), pots, potting soil, sand, compost, a sieve (3 mm mesh), plant labels and a soft-leaded pencil. Because the cuttings root very quickly, rooting hormone is unnecessary.

Vegetative propagation

Offsets and stolons

Offsets form naturally at the base of most garden plants and, in the case of gasterias, consist of plantlets developed from the base of a stem. These usually increase rapidly in size to form dense clusters. An exception is *Gasteria pillansii,* which proliferates from subterranean, horizontal stolons (modified subterranean stems), eventually forming a dense mat-like growth of plants, as opposed to a cluster. A stolon is very similar to an offset, the only difference being that it is carried on a lengthened subterranean stem. The plantlets arising from the base of a leaf cutting are also called offsets.

Division is the easiest and the most common method of propagation for gasterias. Plants can be divided at any time of the year, but the summer months are recommended. Lift a dense gasteria cluster (mother plant and mature offsets) and wash off the soil. Divide the plant by hand or with a sharp knife, taking care not to damage the mother plant. Prune away unnecessary roots and repot in individual containers.

To plant out the offset, fill a small pot with the general potting mixture, firming the soil down. Use a dibber to make a small hole in the centre of the soil, place the new plantlet in the hole and firm the soil down gently around it. (The growth of young plants from a cutting is always much faster than that of plants grown from seeds, as the former are able to draw food from the mother leaf.)

Growing gasterias from cuttings

Leaf cuttings may be taken at any time of year although rooting takes place more quickly in the summer months. When taking leaf cuttings, the lower leaves may be removed without damaging the plant or causing it to look unsightly. Choose a healthy mature plant. Take the lower leaf securely in the right hand while holding the plant with the left. Twist the lower leaf sideways and it should detach itself at the base with minimal damage to the leaf or the plant. To prevent any fungi from entering the open wound at the base of the leaf, dip it into sulphur and leave the detached leaf to heal for two weeks. (Leaf cuttings can also be taken with a knife.)

Prepare the soil medium as for all other plantings. Small containers or shallow seed trays can be used depending on the size of the leaves. Firm down the soil to within 10 mm from the top of the container and place a leaf on the soil in a horizontal position, with the base lightly covered with soil. Keep it moist. When the leaves become turgid and firm, it is a sign that rooting has taken place (usually within a few weeks). The cuttings will soon proliferate from the base to form dense clusters of offsets which may be planted out after a year or two.

When the offsets are ready for planting out, remove the rooted leaf cutting complete with its offsets. Gently wash away the soil from the roots and, using a sharp knife, cut off the offsets near the base of the attachment. If, however, some of the offsets are still very small, do not separate them from the original leaf cutting but replant them all in the original container. Plant out the offsets as described above.

Propagation by seed

Propagation by seed has the advantage of producing variable offspring, of which some individuals may have rewarding horticultural qualities such as attractive leaf colouring.

Pollination and seed production

Most gasterias flower during late winter and spring, producing an elongated inflorescence. Note that the flowers gradually open from below. Gasterias are self-sterile (self-incompatible) and pollen from a different clone (genetically different but belonging to the same species) must be used to achieve success. Gasteria flowers are protandrous, that is, the anthers ripen first, followed within a day or two by the stigma. The yellow pollen is clearly visible on the open anthers (usually situated at the open end of the flower), and it can be collected on a sharpened matchstick. The pollen can then be transferred to the ripe, slightly protruding stigma of a suitable flower of a different clone, but within the same species. After a day, the stigma of the first plant lengthens, the apex moistens and it too is ready for pollination. This is again performed by rubbing pollen from another plant on to the stigma. Tag each flower (tiny jewellery tags are ideal) with details of the parent plants and the date of pollination. Fortunately, when a flower is not pollinated it will abort (drop off), and it is therefore easy to determine whether pollination has been successful.

Once a flower has been pollinated, its ovary swells, the pedicel hardens and the old flower bud is lifted into an erect position. The seeds take about one month to six weeks to develop and when the lowest capsule dehisces, the inflorescence can be cut at the base with secateurs. To collect the seeds, place the inflorescence in a dry paper bag and leave it until all the capsules have opened. Alternatively, leave the inflorescence on the plant and wrap it in a piece of old stocking to prevent the seeds from falling out when ripe. Cut the inflorescence only when all the capsules have dehisced. Store the seeds in a dry place until you are ready to sow them. You can lengthen their viability by keeping them in cold storage, but do not freeze them.

Gasteria flowers can be pollinated by collecting the pollen from the anther of one plant on a matchstick and transferring it to the ripe stigma of another.

Sowing the seeds

The viability of gasteria seeds reduces rapidly after a year. It is therefore advisable to sow seeds before they age. Although they can be sown at any time of year, a higher percentage germinate in the warmer spring, summer and autumn months.

Standard shallow seed trays are recommended, but other smaller (8-cm) pots or containers with deeper bottoms can also be used. The small multi-cavity (six- or eight-cavity) disposable plastic trays are ideal for small quantities of seed. The soil mixture should be sandy, well drained (the mixture recommended under 'Indoor cultivation' is ideal) and preferably sterilized to avoid attack by soil-borne fungal diseases.

Place a sheet of newspaper at the bottom of the seed tray or container to prevent the potting mix from falling out. Fill to about 10 mm from the top and firm down with a pressure board or by hand. Next, wet the soil thoroughly to ensure that the seeds stick to the soil when they are covered with sand, then sprinkle the seeds evenly over the moist soil and cover them with a thin, 2- to 3-mm layer of sand. (Take care not to sow the seeds too densely as fungal soft rot spreads rapidly.) Moisten the soil again, using a watering can with a fine rose. To prevent coarse water droplets from uncovering the seed, start pouring the water away from the tray, and slowly direct it over the tray only when it is flowing in a fine spray. Alternatively, water from below by placing the seed tray in a shallow, water-filled container so that water is drawn up into the soil by capillary action.

The seed tray must not be exposed to full sun, but should be kept in a shady yet warm area. A cold frame, consisting of a wooden box with a sliding glass or fibreglass lid, preferably sloping to one side, is ideal as it retains humidity and reduces temperature fluctuation. Germination usually takes place after approximately three weeks when the first tiny leaves will appear above the sand. At this stage the seedlings are very susceptible to damping off, but this can be prevented by adding a fungicide regularly to the watering.

During the first year the seedlings will require more water than the adult plants, but watering should be reduced during the cooler winter months. The seedlings grow slowly and after a year can be planted out, using a dibber, into small individual containers; alternatively, simply grow them in rows in one large container. Water well once the container or tray has been filled. The best time to transplant gasterias is during the late winter or spring and thereafter feed them bi-monthly with a

When transplanting a young gasteria plant make a hole in the soil of the container, set the plant in it and cover the roots, firming the soil down around them.

Snout beetles *(Brachycercis* sp.) feed on the leaf tissue of gasterias and lay their eggs in the centre of the rosette. The larvae eat into the stem, with the result that the rosette falls apart or rots.

liquid organic fertilizer. Flowering usually occurs after three years, although it can take longer.

Breeding and hybridization

A hybrid is the result of the crossing of two compatible parent plants where the offspring will bear characteristics of both parents. *Gasteria* species hybridize with ease among each other, but can also be hybridized with species of *Aloe*, *Astroloba* and *Haworthia*. Only very few natural hybrids have been found, due to the fact that no two species share the same habitat, and only where habitats meet, have hybrids been found. Like their close relatives, the aloes, hybridization in gasterias usually results in hybrid vigour and the new plants usually grow much faster and flower more profusely than the parents.

New cultivars can be developed as a result of hybridization within a single species or between several *Gasteria* species and are selected on the basis of foliage and/or flowers. Early experiments in hybridization have so far had some promising results, especially from cross-fertilizing such species as *G. rawlinsonii*, *G. carinata*, *G. vlokii*, *G. excelsa* and *G. pulchra*, but most particularly *G. rawlinsonii*.

In their natural habitat all gasterias are pollinated by sunbirds, and when a number of species grow together, sunbirds will visit various plants haphazardly, resulting in seed of several possible combinations. If you wish to try hybridization, first visualize the results you are aiming for – such as a striking leaf colouring or a very floriferous plant – and then choose the parents with care (although sometimes the unexpected can also be rewarding). Once the parents have been selected, net each inflorescence to prevent any possible interferences from pollinators.

When hybridizing two species, select the plants you want to hybridize as well as the choice of 'female' (pollen receiver) and 'male' (pollen donor). Label the 'female' flowers that are ready for pollination with a small tag on which is written the name of the plant, with the name of the 'male' plant below it. Then pollinate as described above.

PESTS AND DISEASES

Hygiene

Small gasteria collections are relatively disease free. However, disease and insects can cause losses, especially in large collections where many species are grown together. Prevention is therefore better than cure and by maintaining a high standard of hygiene, such as cleaning and disinfecting tools and old containers, many diseases and pests can be avoided. Keep your work bench clean, and from time to time wash the floor and work bench with a disinfectant such as Jeyes Fluid. It is also sound practice to use sterilized soil instead of re-using old potting soil. Remove old and sickly plants from your collection and burn them, as adding them to the compost heap can result in the infection of healthy plants at a later stage. Act immediately at the first sign of fungus and remember that ants are often the forerunners of aphids and woolly aphids. Regular inspections of your plants are essential.

Disorders caused by incorrect gardening practices

Gasterias are adaptable to a wide range of conditions, and if a plant is well fed and cared for, little should go wrong. Nonetheless, at times plants do show signs of environmentally induced damage, and gardeners should be aware of the causes of such damage.

Although gasterias will tolerate low light intensity for a considerable time, too little light will result in an enlarged plant with yellowish leaves that are loose and break away easily from the base. Rectify the condition by placing the plant closer to a window or in another suitable spot. Too much light, on the other hand, causes leaf burn, and plants so affected should be moved into a shady position.

Overwatered plants tend to rot, so ensure that the soil is well drained. Too little water will cause the leaves to become deeply channelled.

Severe frost will kill gasterias and exposure to such cold conditions must be avoided by bringing the plants indoors.

Gasterias tolerate a wide range of soil conditions, but most species prefer a light acidic soil with plenty of organic matter. If the soil is too alkaline, add peat or leaf mould. Peaty soils are acid and have the added advantage of preventing fungal diseases. A lack of phosphorus will result in a pale yellowish leaf colour.

Insects and mites

*Snout beetles (*Brachycercis* spp.)*

These robust, hard-skinned beetles prey on leaf tissue and also lay eggs in the centre of the rosette. The larvae eat into the centre of the stem, with the result that the plant rots and the rosette disintegrates. More common on aloes, this pest is only occasionally encountered on gasterias. It is easily controlled with a contact or systemic insecticide which may be dusted or sprayed onto the plant or watered into the soil.

Aphids

Green or black aphids are sometimes a pest on the young inflorescence but are easily controlled by a contact or systemic spray. Woolly aphids are most frequently subterraneous, attaching themselves to roots or to the base of the stem and are thus very difficult to detect. Control by including a systemic or contact poison when watering.

Crown aphids are the most common killers of gasterias. As the name implies, they are found in the crowns of plants, near the base of the leaves in the folds, or between the old leaf bases and they are difficult to detect at an early stage. The first sign of their presence is normally the sighting of small black Argentinian ants which often cover the aphids with sand. The Argentinian ant also spreads the pests to other plants, using the aphids for 'milking' purposes. These aphids multiply rapidly and suck the sap of the leaves to such an extent that the plant disintegrates or rots as a result of secondary fungus infection. Control is easy with a contact or systemic poison which should be added when watering. This will also control the ants.

Scale

Very rarely encountered on gasterias, scale is easily controlled with an oil-based insecticide.

Wasps

I have observed this pest only on wild plants in the Barrydale and Pongola districts. The wasps bore small holes into the leaves and lay their eggs on the inner surface. The larvae hatch, feed and develop inside the leaves until the mature wasps emerge to infest other leaves.

The colourful lily caterpillar (*Brithys* sp.) can cause considerable damage on a gasteria inflorescence.

The aloe gall mite (*Aceria aloinis*) causes distorted growth of leaves or flowers. The enlarged flowers on this inflorescence have been caused by aloe gall mite infestation.

Fruit flies
The first sighting of this common pest on gasteria fruits is when small round holes appear in the young fruiting capsules, indicating that larvae are feeding on the developing seed. Control by watering the plants with a systemic insecticide.

Caterpillars
Although not a common pest, caterpillars have on occasion been observed on gasteria leaves and flowers. They are easily controlled by either spraying or dusting, but it is just as easy, and less damaging to the environment, to remove them by hand.

*Aloe gall mite (*Aceria aloinis*)*
The symptoms of a mite infestation are distorted and monstrous growth on leaves, inflorescences and flowers. The mites are very small and cannot be seen with the naked eye. There are a number of very effective mite control remedies on the market.

Fungi
Damping off (seedlings)
Control and prevent by applying a fungicide when watering.

*Black spot (*Montagnella *sp.)*
Control by regularly applying a contact or systemic fungicide.

Soft or yellow crown rot
This is normally caused by over-watering. Keep watering to a minimum and apply a fungicide.

Fusarium *root rot*
As the name implies, this disease causes the roots to rot, so preventing the uptake of water and nutrients. The first signs are a deteriorating condition of the plant in spite of watering and feeding. It can be prevented by using pre-sterilized soil or by regularly adding a fungicide when watering.

Aloe-ring
Caused by the fungus *Uromyces aloes*, this disease appears as yellowish brown concentric rings on both sides of the leaf surface. It is difficult to control. However, it is more commonly associated with aloes, and gasterias are less frequently affected by it.

Part 2

THE TAXONOMY OF *GASTERIA*

GENERIC TREATMENT

Gasteria Duval: 6 (1809); Haw.: 85 (1812); Haw.: 48 (1819); Haw.: 46 (1821); Haw.: 347 (1827); Bak.: 182 (1880); Bak.: 286 (1896); Berger: 122 (1908); V. Poelln.: 233 (1938a); Jacobsen: 624 (1954); Schelpe: 19 (1958); Jacobsen: 237 (1974); R.A. Dyer: 929 (1976); Rowley: 55 (1976b); Barkhuizen: 186 (1978); Court: 174 (1981); Bond & Goldblatt: 31 (1984). E.J. v. Jaarsveld: 6 (1992). Lectotype: *Gasteria angustifolia* (Ait.) Duv. = *Gasteria disticha* (L.) Haw. [Maire: 71 (1958)].

Aloe section *Curviflorae* Haw.: 12 (1804); Salm-Dyck: 11 (1817); Salm-Dyck: 29 (1836-63).

DESCRIPTION

Slow-growing, perennial succulents, 15-750 mm tall, glabrous in all parts, solitary or proliferating from the base or subterraneous stolons to form dense groups.

Roots succulent, terete, rarely linear-obclavate.

Stem short to acaulescent, decumbent or erect, rarely pendulous and up to 1 m long, branched from the base.

Leaves with dimorphic ontogeny from juvenile to adult stage, at first distichous, often remaining distichous or becoming spirally arranged or rosulate, erectly spreading to recurved, sometimes twisted sideways, lorate, lanceolate, triangular, linear to ensiform, rarely undulating, 15-550(-600) x 7-220 mm, often with the apices incurved (more so during the dry season) or rarely recurved; base amplexicaul; adaxial surface plane to canaliculate, becoming more so during the dry season, light to dark green to rarely glaucous, becoming reddish tinged when exposed to sunlight, usually with dense white spots in transverse bands; abaxial surface convex or plane when distichous, otherwise with a distinct (rarely indistinct) marginiform excentric keel; epidermis asperulous, tuberculate, rugulose or smooth; tubercles white or greenish, shiny, translucent, domed, convex, rarely dense and confluent forming a reticulation, often secondarily tuberculate; margin acute, rounded, rarely truncate, wavy or rugulose; tuberculate-crenulate, tuberculate-denticulate or tuberculate-serrulate, the tubercles merging towards the apex, forming a continuous white margin which may be entire, serrulate, crenulate or rarely denticulate, rarely with a white semi-translucent sharp margin; apex truncate, obtuse, acute or acuminate, rarely retuse, mucronate.

Inflorescence terminal, 1-5 per plant, racemose or paniculate, 0,07-1,9 m tall, erectly spreading to horizontal; scape basally flattened, biconvex, terete towards apex, green with a powdery bloom, often pink in upper half, rarely minutely and sparsely denticulate; bracts persistent, triangular-acuminate, scarious, sometimes piliferous, 5-40 x 2-25 mm; pedicel pendulous, pink, shorter than the perianth, laxly and secundly arranged in the upper half.

Perianth six-lobed, 12-47(-50) mm long, shortly stipitate, the lobes fused into a cylindrical curved tube, basally gasteriform for up to two-thirds of the perianth length; gasteriform portion globose, globose-elliptic or narrow-elliptic, usually constricted below the gasteriform portion; the latter sometimes indistinct or similar to upper portion, pink to reddish pink, rarely nearly white, becoming white in upper half with green striations; apices free, erect, erectly spreading to recurved, obtuse, light pink to white, rarely yellowish, with green median stripes; three inner segments dorsally fused to three outer segments, margins free, membranous and canaliculate, rarely constricted just below the apex of the ovary, the membranous margins adpressed against the ovary, filled with nectar; apices of inner segments slightly longer and more recurved than those of the outer segments.

Stamens six, 15-35 mm long, included or rarely exserted up to 4 mm, often three lengthening in advance; filaments white, slightly flattened, becoming contracted after anthesis; anthers dorsifixed, versatile and introrse, dorsally grouped together in the throat, 2,2-3,5 mm long, oblong-ovoid, bilocular, longitudinally dehiscing; pollen yellow.

Ovary 5-8 mm long, oblong-ovoid, three-chambered, six-grooved, green, rarely with three apical appendages; placentation axile, ovules numerous; stigma apical, minute, trilobate, becoming shortly exserted, capitate and exceeding the anthers after anthesis.

Capsule oblong, oblong-ovoid, rarely somewhat clavate, three-angled, 12-43 mm long and 6-18 mm in diameter, shortly stipitate for 2 to 6 mm, green at first, becoming woody and dehiscing longitudinally; outer walls with transverse ridges; apex truncate, with the apices somewhat recurved.

Seeds black, 2-8 mm long, compressed, irregularly angled and obscurely oblong in outline.

KEY TO THE SPECIES BASED ON COMPLETE MATERIAL

1a Perianth 35-50 mm long; gasteriform portion and tube about the same diameter:
- **2a** Leaves distichous; epidermis asperulous (Namaqualand) (10 *G. pillansii*):
  - **3a** Leaves longer than 70 mm; stamens usually not exserted ... 10a *G. pillansii* var. *pillansii*
  - **3b** Leaves shorter than 70 mm; stamens shortly exserted ... 10b *G. pillansii* var. *ernesti-ruschii*
- **2b** Leaves rosulate; epidermis smooth or tuberculate:
  - **4a** Leaves distinctly rugulose-tuberculate (E Tvl, Zululand) ... 1 *G. batesiana*
  - **4b** Leaves smooth:
    - **5a** Capsule 35-43 mm long, seeds 6-8 mm in diam. (S E Cape coast) ... 3 *G. acinacifolia*
    - **5b** Capsule 18-25 mm long, seeds 3-4 mm in diam. (Natal) ... 2 *G. croucheri*

1b Perianth shorter than 35 mm; gasteriform portion distinct; diameter of tube smaller than the diameter of the gasteriform portion:
- **6a** Plants caulescent (with a short to long stem):
  - **7a** Plants decumbent or erect with short stem (13 *G. bicolor*):
    - **8a** Leaves shorter than 90 mm (Grahamstown distr.) ... 13b *G. bicolor* var. *liliputana*
    - **8b** Leaves longer than 90 mm (widespread in E Cape) ... 13a *G. bicolor* var. *bicolor*
  - **7b** Plants with long pendulous stems (Baviaanskloof) ... 16 *G. rawlinsonii*
- **6b** Plants acaulescent:
  - **9a** Leaves distichous, remaining distichous:
    - **10a** Leaf epidermis smooth (Little Karoo, Willowmore) ... 12 *G. brachyphylla*
    - **10b** Leaf epidermis asperulous or tuberculate:
      - **11a** Leaf epidermis asperulous (Breede River valley, S W Great Karoo) ... 11 *G. disticha*
      - **11b** Leaf epidermis tuberculate:
        - **12a** Apex retuse (Worcester) ... 9b *G. carinata* var. *retusa*
        - **12b** Apex obtuse or acute:
          - **13a** Perianth bright reddish pink (Gamtoos):
            - **14a** Leaves dark green (tuberculate) ... 6b *G. nitida* var. *armstrongii*
            - **14b** Leaves glaucous (asperulous-tuberculate) ... 15 *G. glomerata*
          - **13b** Perianth tricoloured:
            - **15a** Perianth 14-16 mm long (Suurberg) ... 14 *G. baylissiana*
            - **15b** Perianth 18-25 mm long (Lower Breede River) ... 9c *G. carinata* var. *verrucosa*
  - **9b** Leaves rosulate, keeled:
    - **16a** Leaf epidermis distinctly tuberculate:
      - **17a** Outer leaves recurved, retuse ... 7 *G. ellaphieae*
      - **17b** Outer leaves spreading, not retuse ... 9a *G. carinata* var. *carinata*
    - **16b** Leaf epidermis smooth or asperulous:
      - **18a** Leaf epidermis asperulous (Swartberg) ... 8 *G. vlokii*
      - **18b** Leaf epidermis smooth:
        - **19a** Leaves linear triangular ... 5 *G. pulchra*
        - **19b** Leaves triangular, triangular-lanceolate or lanceolate:
          - **20a** Leaves triangular to lanceolate; margin sharply serrulate (E Cape) ... 4 *G. excelsa*
          - **20b** Leaves triangular to triangular-lanceolate; margin entire or tuberculate:
            - **21a** Perianth bright reddish pink (S E Cape), apices yellowish, margin entire ... 6a *G. nitida* var. *nitida*
            - **21b** Perianth tricoloured, apices green; margin tuberculate (S Cape) ... 9a *G. carinata* var. *carinata*

KEY TO THE SPECIES BASED ON VEGETATIVE MATERIAL

1a Plants acaulescent (without stem):
- **2a** Leaves rosulate, tapering, distinctly keeled:
  - **3a** Leaf epidermis tuberculate or asperulous:
    - **4a** Leaf epidermis asperulous (Swartberg) .... 8 *G. vlokii*
    - **4b** Leaf epidermis tuberculate:
      - **5a** Leaf epidermis rugulose-tuberculate (E Tvl, Swaziland, N Natal) .... 1 *G. batesiana*
      - **5b** Leaf epidermis not rugulose-tuberculate:
        - **6a** Leaf epidermis uniformly but sparsely tuberculate (S Cape) .... 9a *G. carinata* var. *carinata*
        - **6b** Leaf epidermis densely domed tuberculate (Kouga region) .... 7 *G. ellaphieae*
  - **3b** Leaf epidermis smooth:
    - **7a** Leaves linear triangular (Lower Gamtoos) .... 5 *G. pulchra*
    - **7b** Leaves triangular to triangular-lanceolate:
      - **8a** Margin serrulate:
        - **9a** Margin not sharply serrulate (Natal, E and S Cape coast):
          - **10a** Leaves erectly spreading, green (E and S Cape coast) .... 3 *G. acinacifolia*
          - **10b** Leaves spreading, often glaucous .... 2 *G. croucheri*
        - **9b** Margin sharply serrulate (Transkei, E Cape) .... 4 *G. excelsa*
      - **8b** Margin entire (E Cape) .... 6a *G. nitida* var. *nitida*
- **2b** Leaves distichous, lorate, not keeled:
  - **11a** Leaf surface smooth (Little Karoo) .... 12 *G. brachyphylla*
  - **11b** Leaf surface asperulous or tuberculate:
    - **12a** Leaf surface asperulous:
      - **13a** Margin wavy (Worcester, Lower Great Karoo) .... 11 *G. disticha*
      - **13b** Margin straight (Namaqualand) .... 10 *G. pillansii*
    - **12b** Leaf surface tuberculate or asperulous-tuberculate:
      - **14a** Leaves patent or recurved (Kouga and Gamtoos):
        - **15a** Leaves dark green, tuberculate (Lower Gamtoos) .... 6b *G. nitida* var. *armstrongii*
        - **15b** Leaves glaucous green, asperulous-tuberculate (Kouga) .... 15 *G. glomerata*
      - **14b** Leaves erectly spreading:
        - **16a** Margin wavy (Worcester, Robertson) .... 9b *G. carinata* var. *retusa*
        - **16b** Margin straight:
          - **17a** Apex truncate (Grahamstown) .... 14 *G. baylissiana*
          - **17b** Apex obtuse or acute (S Cape) .... 9c *G. carinata* var. *verrucosa*

1b Plants caulescent (with short to long leafy stem):
- **18a** Plants decumbent, leaf epidermis smooth (widespread, E Cape) (13 *G. bicolor*):
  - **19a** Leaves shorter than 90 mm (Grahamstown district) .... 13b *G. bicolor* var. *liliputana*
  - **19b** Leaves longer than 90 mm (widespread, E Cape) .... 13a *G. bicolor* var. *bicolor*
- **18b** Plants pendulous, leaf epidermis asperulous (Baviaanskloof) .... 16 *G. rawlinsonii*

Gasteria batesiana may be found on sheer cliff faces overlooking the Pongola River near the Klipwal gold mine. *Aloe rupestris* (foreground) also grows in the area.

1. **Gasteria batesiana** *Rowley*, in Jacobsen & Rowley: 32 (1955); Schelpe: 20 (1958); Jacobsen: 502 (1960); E.J. v. Jaarsveld: 8 (1992). Neotype: *Rowley s.n.* (RNG, neo.!).

HISTORICAL BACKGROUND

Gasteria batesiana is one of the most striking *Gasteria* species, hence its popularity in cultivation in South Africa and abroad. It was first collected in May 1924 by Frank Frith, the first horticulturist of the South African Railways, and a keen succulent collector and explorer who may have introduced *G. batesiana* to Europe. Although it had been in cultivation in Europe and South Africa since the early part of this century, its location in the veld remained unknown until 1958 when Theo Sprengel of Ingogo, Natal, collected living specimens near Piet Retief in the eastern Transvaal and gave them to Harry Hall at Kirstenbosch.

The specific epithet commemorates John Bates, a trolley-bus conductor in London and a keen collector of South African succulents (Jacobsen & Rowley 1955).

DESCRIPTION

Plants acaulescent, decumbent to erect, 30-100 mm tall and 80-300 mm in diameter, proliferating from the base to form small to large groups, rarely solitary. *Roots* succulent, terete. *Leaves* distichous at first, becoming rosulate, 50-180 mm long, 15-40 mm broad at the base, firm but brittle, triangular, triangular-lanceolate or linear, erectly spreading or becoming recurved, adaxial surface channelled, plane towards apex; abaxial surface convex, with an excentric marginiform keel; both surfaces dark green with dense white spots arranged in transverse bands, distinctly and densely rugulose-tuberculate, with larger and smaller green or white tubercles; the tubercles conical or subconical, often confluent forming ridges; margin cartilaginous, serrulate, rarely denticulate; apex acute, rarely obtuse, mucronate. Juvenile leaves lorate, densely tuberculate, apex obtuse, mucronate. *Inflorescence* racemose, erectly spreading and curved, 300-450 mm long; bracts 6-12 mm long, 2-5 mm broad at the base; pedicel 9 mm long. *Perianth* 35-40 mm long, stipitate for 3-5 mm, gasteriform basally (narrow-elliptic) for half its length; the gasteriform portion 6-9 mm in diameter (often triangular in cross section), light pink, upper half white with green striations, inflated to the same diameter as lower portion (with a slight constriction in the middle); apices erect becoming erectly spreading, obtuse, white with green median stripes; margins of inner segments free, canaliculate for the length of the ovary. *Stamens* 30-35 mm long; anthers 2,5 mm long, included or shortly exserted. *Ovary* 7 mm long, 3 mm in diameter; style 15 mm long, stigma included or exserted for up to 5 mm, curved upwards. *Capsule* 16-20 mm long. *Seeds* 4-6 mm long, 2-3 mm wide. *Flowering time* from October to December, but occasionally throughout the year.

DISTRIBUTION AND HABITAT

Gasteria batesiana is the most northerly species in the genus, occurring from northern Natal to the Olifants River valley in the north-eastern Transvaal. It is found in the Lowveld at elevations from 500 to 700 m, in hot, dry, frost-free river valleys in mountainous terrain. The habitat consists of shady southern and eastern aspects of wooded rocky ridges and krantzes, where it occurs in shallow, humus-rich soil. It has been recorded as occurring on rhyolite, sandstones and shales, quartzitic rocks, and dolomites. Near the Klipwal gold mine in the Pongola district, plants are found on a sheer south-facing rock face overlooking the river. They abound here in shallow pockets in quartzitic sandstone together with other xerophytic species, the vegetation forming a mosaic of valley bushveld and lowveld. The plants proliferate to form small, dense groups.

DIAGNOSTIC FEATURES AND VARIATIONS

Gasteria batesiana is characterized by its small rosette of firm, brittle, triangular to linear, spreading to recurved, distinctly rugulose-tuberculate leaves, and its simple, racemose inflorescence bearing large flowers of 35-38 mm in length. The leaves are often very dark, almost a blackish green. It is closely related to *G. croucheri*, a much larger sister species (perianth similar) with smooth, often glaucous leaves and a paniculate inflorescence.

In vegetative features *G. batesiana* closely resembles *G. carinata* var. *carinata* and *G. ellaphieae* from the southern and south-eastern Cape; these, however, have a much smaller and different perianth. *G. carinata* has leaves that are sparingly tuberculate (rarely smooth) and *G. ellaphieae* has leaves with dense, domed tubercles. *G. batesiana* is variable in the size, colour and shape of the leaves. A collection by Gerstner from the Little Flower Mission at Eshowe has smooth leaves and requires further study. It appears to be intermediate between *G. batesiana* and *G. croucheri*, corroborating their close affinity.

CULTIVARS

'Pongola' The attractive form so common in cultivation is thought to have come from the Pongola valley. The plants are most often solitary with very striking, triangular, often recurved, mottled, dark green leaves. They flower during September and October.

'Golela' This form, another from the Pongola valley, has linear leaves. It was collected at the Pongola dam and brought into cultivation by Dave Hardy.

'Barberton' This cultivated form from Barberton has very dark, blackish green leaves with an obtuse apex.

'Penge' Plants from the Penge district have attractive, mottled, linear-lorate, almost terete, leaves. They occur in very shallow dolomite soils.

'Klipwal' A large form that originates from sheer cliff faces near the Klipwal gold mine (the largest recorded for *Gasteria batesiana*) has leaves up to 260 mm long and about 70 mm broad at the base. The perianth is also larger than normal, measuring 45 mm in length.

VOUCHERS

Gerstner 1210/35 (BOL, NBG); *Hardy 1550* (PRE, NBG, K); *Strey 5292* (PRE); *Thompson 8289* (PRE); *Van Jaarsveld 9132* (NBG).

Gasteria batesiana (all parts life size): 1. mature plant with juvenile; 2. raceme; 3. infructescence; 4. capsule and seeds.

The valley of the Mtentu River, Transkei, is one in which *Gasteria croucheri* occurs. The plants may be found on rocky ridges and cliff faces.

2. **Gasteria croucheri** *(Hook. f.) Bak.*: 196 (1880); Bak.: 301 (1896); Berger: 157 (1908); V. Poelln.: 238 (1938a); Jacobsen: 630 (1954); E.J. v. Jaarsveld: 8 (1992). *Aloe croucheri* Hook f.: t. 5812 (1869). Type: South Africa, *T. Cooper s.n.* (K, holo.!).

HISTORICAL BACKGROUND

Gasteria croucheri was first collected and introduced to cultivation by Thomas Cooper in 1860.

G. croucheri is highly prized by Zulu warriors for its magical properties, as it is believed to render the user partly invisible. It is frequently used during faction fights. This has led to stripping of the veld and this species has become very rare in Natal.

DESCRIPTION

Plants acaulescent, decumbent to erect, 250-400 mm tall and up to 600 mm in diameter, solitary, dividing or proliferating from the base to form dense groups. *Roots* succulent, up to 6 mm in diameter. *Leaves* rosulate, 200-360 mm long, 30-100 mm broad at the base, triangular, linear-lanceolate, rarely lorate, erectly spreading, rarely patent or recurved; adaxial surface canaliculate, plane towards apex; abaxial surface somewhat convex with a distinctly excentric keel; both surfaces dark green, often glaucous, with dense white spots arranged in transverse bands; epidermis smooth, rarely slightly asperulous; margin tuberculate-serrulate, rarely denticulate; apex obtuse or acute, mucronate. Juvenile leaves distichous, lorate, patent to erectly spreading, smooth or asperulous; apex acute, rarely obtuse, mucronate. *Inflorescence* variable: a flat-topped panicle up to 500 mm tall, or racemose, with or without a pair of side branches; racemes horizontal (rarely erectly spreading), often drooping; scape 6-10 mm broad at the base, flattened; floral bracts 5 mm long, piliferous; pedicel 6-7 mm long, pink. *Perianth* 28-40 mm long, stipitate for up to 6 mm, gasteriform basally (narrow-elliptic) for half its length; the gasteriform portion 5-9 mm wide (often triangular in cross section), pink, the upper half white with green striations, inflated to the same diameter as the lower portion (with a slight constriction in the middle); apices erect becoming erectly spreading, obtuse; margins of inner segments free and channelled at the base for 5 to 6 mm, diverging gradually towards the apex. *Stamens* 28-30 mm long; anthers 2 mm long, included or shortly exserted. *Ovary* 7-12 mm long and 3 mm in diameter; style 20-25 mm long; stigma included or shortly exserted, curved upwards, minute. *Capsule* 18-25 mm long, obtuse at apex. *Seeds* 3-4 mm long, 2-3 mm wide. *Flowering time* November to February.

DISTRIBUTION AND HABITAT

Gasteria croucheri is confined to Natal and the northern Transkei region, from the Msikaba River in the south to Mkuze in the north. Although not common, this species is widely distributed, from close to sea-level to an altitude of 1 000 m. It is found in dry, hot, rocky and wooded river valleys, and is almost always associated with quartztitic sandstone. The habitat usually consists of sheer cliffs and the plants occur in the shade of the northern variation of valley bushveld. They are also found on rocks in coastal forest and thornveld. The soil is shallow and humus-rich, and the plants seem to be confined to southern and eastern aspects. Precipitation is mainly in summer and ranges from 500 to 1 000 mm per annum. The climate is hot during summer and mild in winter. Populations of this species at Oribi Gorge (southern Natal) occur on sheer cliffs, where they grow in the shade of short xerophytic trees and shrubs.

DIAGNOSTIC FEATURES AND VARIATIONS

Gasteria croucheri is a very variable but distinct species, characterized by its large perianth 35-40 mm long, inflated in the lower and upper half, and slightly constricted in the middle. It is further characterized by its rosette of spreading, triangular-lanceolate or triangular (rarely linear), often glaucous leaves. The juvenile leaves may be asperulous or smooth. Its inflorescence and habit vary: in the north the plants divide to form dense clusters and the inflorescence consists of a simple raceme which may sometimes bear a pair of side branches. Here the leaves are linear and often droop from the sheer cliff faces. In the south (Port Shepstone and northern Transkei), the plants are solitary and the inflorescence consists of a spreading, flat-topped panicle, often with drooping racemes. In floral characters it is closest to *G. batesiana* and *G. acinacifolia*, the former a smaller sister species (perianth similar) with a distinctly rugulose-tuberculate leaf epidermis and a simple raceme. *G. acinacifolia* is a larger species with taller panicles and larger bracts (20 mm long). It is also clearly distinguished by its tuberculate juvenile leaves, larger capsules (30-43 mm) and seeds of up to 8 mm in diameter. *G. croucheri* can also be confused with *G. excelsa*, which is also a large robust species, but the latter has a smaller perianth that is distinctly gasteriform basally for more than half of its length.

CULTIVARS

Gasteria croucheri, being variable, has several local forms of which the following have been brought into cultivation:

'Oribi' In the Oribi Gorge region (Port Shepstone district), plants have spreading, triangular to triangular-lanceolate leaves. The paniculate inflorescence has a flat top and the spreading racemes are often drooping. The juvenile leaves are smooth.

'Shongweni' In Durban and towards Zululand the plants become smaller with linear, lanceolate leaves which are somewhat glaucous. The inflorescence is a simple raceme, occasionally with a pair of side branches. The plants divide to form dense clusters. Flowering occurs from October to January.

'Umgeni' A very interesting form from the Umgeni valley where the plants divide to form dense clusters. The long, linear-lanceolate leaves (up to 470 mm long and 45 mm broad at the base) droop over the sheer cliff faces where it occurs.

VOUCHERS

Nichols 609 (NBG); *Prosser 1411* (NH); *Strey 9456* (PRE, K); *Van Jaarsveld 2195* (NBG); *Ward 3330* (PRE).

Gasteria croucheri (all parts life size): 1. mature plant; 2. perianth segments and stamens; 3. gynoecium and stipe; 4. cross section through leaf; 5. juvenile plant; 6. part of inflorescence.

Gasteria acinacifolia growing in dune thicket at the Maitland River mouth, near Port Elizabeth, in association with the reddish groundcover, *Crassula expansa*.

3. **Gasteria acinacifolia** *(Jacq.) Haw.*: 49 (1819); Haw.: 359 (1827); Bak.: 196 (1880); Bak.: 300 (1896); Berger: 155 (1908); V. Poelln.: 235 (1938a); Jacobsen: 625 (1954); E.J. v. Jaarsveld: 8 (1992). Iconotype: Jacq., *Eclog. pl. rar.* 49 t. 31 (1811-16).

HISTORICAL BACKGROUND

It is not known who first collected *Gasteria acinacifolia*, but according to Jacquin (1811-16), it was growing in Dutch gardens in 1809. It was later reintroduced to European gardens by Bowie and Ecklon (Berger 1908).

DESCRIPTION

Plants acaulescent, decumbent to erect, 250-750 mm tall and up to 650 mm in diameter, solitary, or dividing or proliferating from the base to form small groups. *Roots* succulent, up to 6 mm in diameter. *Leaves* rosulate, 220-600 mm long, 45-100 mm broad at the base, linear-lanceolate to lorate, erectly spreading and sometimes falcate (rarely patent or becoming recurved); adaxial surface canaliculate, plane towards apex; abaxial surface somewhat convex with a distinctly excentric keel; both surfaces dark green, with dense white spots arranged in transverse bands; epidermis smooth, rarely slightly tuberculate; margin cartilaginous-serrulate, rarely entire; apex acute, rarely obtuse, mucronate. Juvenile leaves distichous, lorate, patent to erectly spreading, with distinct domed tubercles, rarely smooth; apex acute or obtuse, mucronate. *Inflorescence* racemose, variable, usually a flat-topped panicle up to 1 m tall; side branches horizontal to erectly spreading, rarely a simple raceme with two spreading side branches; scape 6-25 mm broad at the base, flattened; bracts 18-40 mm long, 6-14 mm broad at the base; floral bracts as long as the pedicels, piliferous; pedicel 10-25 mm long, pink. *Perianth* 35-45(-50) mm long, stipitate for up to 5 mm, gasteriform basally (narrow-elliptic) for half its length; the gasteriform portion 5-9 mm wide (often not constricted and indistinct from lower portion), pink, the upper half white with green striations; apices erect becoming erectly spreading, obtuse; margins of inner segments free and channelled at the base for 5-6 mm, the margins diverging gradually towards the apex. *Stamens* 28-40 mm long; anthers 3,5 mm long, included. *Ovary* 10-12 mm long, 3 mm in diameter; style 27-29 mm long, included or exserted for up to 5 mm; stigma curved upwards, minute. *Capsule* 35-43 mm long, truncate or obtuse at apex. *Seeds* 6-8 mm long, 5-6 mm wide. *Flowering time* September to December.

DISTRIBUTION AND HABITAT

Gasteria acinacifolia is confined to a narrow belt along the southern Cape coast, from Knysna in the south to Port Alfred in the north-east. It is a strictly coastal species and is found scattered on well-vegetated sand-dunes or rocky hills close to the shore. Plants usually grow in the shade of coastal dune thicket. The littoral soils are sandy, humus-rich and well drained. The area has a rainfall of 300-800 mm per annum, which is spread throughout the year. The summers are hot and the winters mild and frost-free.

G. acinacifolia at the Maitland River mouth occurs on the eastern aspect of coastal sand-dunes, in shade as well as full sun. The vegetation is shrubby and up to 2 m tall.

At Knysna – the most westerly limit of the species range – plants can be seen on a steep, south-facing, rocky embankment (quartzitic sandstone) close to the shore, growing amongst coastal scrub.

DIAGNOSTIC FEATURES AND VARIATIONS

Gasteria acinacifolia is characterized by its large rosette of smooth, erectly spreading, linear to linear-lanceolate leaves and its large perianth, which is 35-50 mm long and indistinctly gasteriform basally for half of its length. The juvenile leaves are usually distinctly tuberculate. It is at once distinguished from any other *Gasteria* species by its 35-43 mm long capsule, which is truncate at the top, and its 6-8 mm long seeds. Regarding these two characteristics it is larger than any other gasteria. It is further characterized by its large, flat-topped panicles. It is fairly widely distributed and, together with *G. croucheri* and *G. batesiana*, seems to be among the most primitive species in the genus. Its flowers are similar to those of these two species. The perianth of *G. acinacifolia* is usually not constricted in the middle (or if constricted, not markedly so), and the diameter of the upper portion is slightly less than that of the lower. *G. batesiana* is a smaller species with a distinctly rugulose-tuberculate leaf epidermis and a simple raceme. *G. acinacifolia* is distinguished from *G. croucheri* by its distinctly tuberculate juvenile leaves, taller inflorescence and large capsules and seeds. When not in flower, *G. acinacifolia* may be confused with *G. excelsa*, which is also a large robust species, but which has a smaller perianth, distinctly gasteriform basally for about two thirds of its length. The name *G. acinacifolia* has been wrongly applied in the past to large forms of *G. pulchra* of the Gamtoos valley. The latter is a smaller species with linear-triangular to ensiform, channelled leaves (apex acute), with a smaller perianth.

CULTIVARS

As *Gasteria acinacifolia* is variable, the following local forms have been brought into cultivation:

'Knysna' Plants from Knysna are smaller with spreading to recurved, linear-lanceolate leaves and a simple, spreading racemose inflorescence, occasionally with a pair of side branches. The juvenile leaves are tuberculate. This cultivar flowers during December.

'Tsitsikamma' This form from Plettenberg Bay and Nature's Valley is large with long, erectly spreading, linear-lanceolate, falcate leaves and a lax flowering panicle with erectly spreading side branches. The juvenile leaves are smooth or only slightly tuberculate. This plant flowers during November.

'Algoa' Plants from the Port Elizabeth district are usually solitary, robust and large, exceeding any other gasteria in size (without inflorescence). The leaves are erectly spreading and lanceolate to triangular-lanceolate, and the paniculate inflorescence has a characteristic flat top. This cultivar flowers during September.

VOUCHERS

Fourcade 2365 (STE, K); *Marloth 12602* (PRE); *Patterson 2166* (BOL); *Van Jaarsveld 6891* (NBG); *White 1073* (GRA).

Gasteria acinacifolia (all parts life size): 1. part of raceme; 2. mature leaf; 3. juvenile plant; 4. perianth showing the inner channelled segments; 5. gynoecium and androecium; 6. cross section through leaf; 7. capsule and seed.

Gasteria excelsa at Fraser's Camp, east of Grahamstown, where it shares the habitat with *Euphorbia triangularis*. It grows in full sun or in the shade of subtropical thicket.

4. **Gasteria excelsa** *Bak.*: 195 (1880); Bak.: 299 (1896); Pole Evans: t. 665 (1937). E.J. v. Jaarsveld: 10 (1992). Type: Cape Province, Chalumna River, *T. Cooper s.n.* (K, holo.!).

HISTORICAL BACKGROUND

Gasteria excelsa was first collected by Thomas Cooper in 1860 near the Chalumna River (Baker 1880). He introduced live plants to Europe.

DESCRIPTION

Plants acaulescent, robust, decumbent to erect, solitary, 300-600 mm tall, 600-750 mm in diameter. *Roots* succulent, up to 3 mm in diameter. *Leaves* in a dense rosette, 100-400 mm long, 100-180 mm broad at the base, triangular to triangular-lanceolate, erectly spreading to somewhat recurved, brittle; new plants proliferate from leaf fragments; adaxial surface canaliculate, dark green; abaxial surface somewhat convex with a distinct excentric keel; both surfaces dark green with indistinct white spots arranged in transverse bands, which (rarely) may be barely visible, rarely striate; epidermis smooth; margin often very sharp, white, semi-translucent, cartilaginous, serrulate; apex acute or obtuse, mucronate. Juvenile leaves distichous, densely white-spotted, patent, lorate, tuberculate; tubercles domed, rarely conical, lower surface rarely somewhat asperulous; apex obtuse, mucronate. *Inflorescence* a very large spreading panicle, 1-3 per plant, 1-1,9 x 1,5-2,4 m; scape up to 34 mm in diameter at the base, branched 550-750 mm from the base; side branches erectly spreading, rarely horizontally spreading, bearing 30 to 100 flowers (500 to 2 700 per plant); bracts 15-20(-25) mm long and 15-25(-30) mm broad at the base; floral bracts 7-8 mm long and 2 mm broad at the base; pedicels 7-8 mm long. *Perianth* 22-26 mm long, stipitate for 2-3 mm, gasteriform basally for more than half of its length, thence constricted to a tube 5 mm in diameter; the gasteriform portion narrow-elliptic, light pink to rarely white, 6-7 mm in diameter at its widest point; apices erect, becoming slightly erectly spreading, white with pale green median stripes, obtuse; margin of inner segments free, canaliculate for the length of the gasteriform portion. *Stamens* 20-22 mm long; anthers oblong, 2 mm long. *Ovary* 7 mm long, 2,5 mm in diameter; style 10-14 mm long, included, exceeding anthers after anthesis; stigma becoming capitate after anthesis. *Capsule* 17-20 mm long, 8-12 mm broad. *Seeds* 4-5 mm long and 2-3 mm wide. *Flowering time* November to February.

DISTRIBUTION AND HABITAT

Gasteria excelsa is widely distributed in the eastern Cape and Transkei, from Port Alfred in the south-west to the Kentani district in the north-east. It occurs from near sea-level to an altitude of 1 400 m. The habitat consists of evergreen coastal forest and forested river valleys. *G. excelsa* is generally confined to shady undergrowth. It grows in well-drained soils derived from shales and mudstones, as well as on quartzitic sandstone, and on dolerite. Precipitation ranging between 600 and 1 000 mm annually occurs in summer and winter, with winters tending to be dry in the north-east. In some parts of its distribution range, especially where the forests have been stripped for firewood, plants are often eaten by cattle. The leaves are brittle and new plants proliferate from pieces of broken leaves. Among the veld types in which this species occurs are valley bushveld, coastal forest and thornveld. The plants occur abundantly among Mesozoic rocks on well-drained, shady, north-facing banks of rivers.

Near Cala, at the north-eastern end of its distribution range, *G. excelsa* was observed on narrow rocky ledges of sheer sandstone cliffs, on south-eastern and northern slopes, rarely exposed to full sun. The surrounding vegetation consists of highland sourveld. These variations that occur as solitary scattered specimens are somewhat smaller than usual and grow in association with shrubs, small trees and other succulents. This form flowers in November.

DIAGNOSTIC FEATURES AND VARIATIONS

Gasteria excelsa is a variable but distinct species. The very large, spreading, much-branched panicles, 1-1,9 m tall and up to 2,4 m in diameter, distinguish this from any other species. The perianth is light pink and gasteriform for two thirds of its length. Plants of this species are also recognized by their very large acaulescent, often solitary, rosettes of triangular-lanceolate leaves with sharp, serrulate margins; these leaf margins are unique in the genus. *G. excelsa* is closely related to *G. pulchra*, which has a somewhat similar acaulescent rosette and tuberculate juvenile leaves, as well as a paniculate inflorescence. The latter, however, is immediately distinguished from *G. excelsa* by its narrowly triangular leaves and reddish pink perianth. *G. pulchra* also differs in that it flowers in spring.

G. excelsa is also related to *G. nitida*, which differs by its branched, erect to erectly spreading, racemose inflorescence, the bright reddish pink perianth and the entire leaves. *G. excelsa* may also be confused with very large, robust forms of *G. bicolor*, *G. croucheri* and *G. acinacifolia*, but *G. bicolor* always has a short leafy stem and flowers from July to October, while *G. acinacifolia* and *G. croucheri* have flat-topped panicles and a differently shaped, larger perianth, indistinctly gasteriform basally for half its length.

CULTIVARS

The late Frank Stayner, former Curator of the Karoo National Botanical Garden, collected an interesting xeromorphic variation near Cala. It has short leaves with wrinkled margins and pronounced keels. This very attractive form occurs in grassveld in full sun and has been given the cultivar name 'Cala'.

'Gaika' An attractive large form with striations on the leaves; it originates from Gaika Fort near Stutterheim.

'Nqancule' A large form was introduced from Nqancule in the Transkei. The margin of the leaves of this form is extremely sharp and dangerous.

VOUCHERS

Court 284 (NBG); *Dyer 4527* (PRE); *Galpin 8189* (PRE, GRA); *Reynolds 206* (BOL); *Van Jaarsveld & Marthinus 7622* (NBG).

Gasteria excelsa (all parts life size, except 4 and 5): 1. mature plant; 2. part of inflorescence; 3. juvenile plant; 4. perianth segments and stamens (x2); 5. gynoecium (x2); 6. capsule and seed.

Dense subtropical thicket near Hankey, in the eastern Cape, shelters *Gasteria pulchra*.

5. **Gasteria pulchra** *(Ait.) Haw.*: 86 (1812). E.J. v. Jaarsveld: 10 (1992). Iconotype: Miller: t. 292 (1759), the only element cited by Ait. (1789). *Aloe pulchra* Haw. (1804). *Aloe maculata* var. *pulchra* Ait. in *Hortus kewensis* 469 (1789).

HISTORICAL BACKGROUND

It is not known who first collected *Gasteria pulchra*, but according to Miller (1759) it 'was raised from seeds in the Chelsea Gardens (London), which were brought from the Cape of Good Hope by Captain Hutchinson of the *Godolphin Indiaman*'.

One of the paratypes cited by De Candolle includes a plate in Tilli and Florentino's *Catalogus plantarum horti Pisani* (1723). The identity of this illustration is unclear; it appears to be some sort of gasteria with erect flowers.

A fragmentary specimen in the Thunberg Herbarium (*8591a*) possibly represents the above-mentioned taxon, but its identity cannot be established with certainty as it may also be related to narrow-leaved individuals of *G. bicolor*.

DESCRIPTION

Plants acaulescent, decumbent to erect, 200-360 mm tall and of the same diameter, solitary or proliferating from the base to form small groups. *Roots* succulent, up to 3 mm in diameter. *Leaves* distichous at first, becoming rosulate, rarely remaining distichous, 240-360 mm tall and 25-40 mm broad at the base, erectly spreading, often falcate, linear-ensiform to linear-acuminate, tapering, semi-terete to triangular in cross section; adaxial surface distinctly channelled, becoming more so during the dry season, plane towards apex; abaxial surface somewhat convex, with an indistinct excentric keel; both surfaces dark green with dense white spots, arranged in transverse bands; epidermis smooth, rarely slightly tuberculate; margin cartilaginous, serrulate; apex acute or acuminate, mucronate. Juvenile leaves distichous, erectly spreading, lorate, acuminate, distinctly tuberculate, mucronate. *Inflorescence* an erect to erectly spreading lax panicle 0,35-1,50 m tall, rarely simple when plants are small and later with a pair of erectly spreading side branches; scape 5-10 mm at the base; branched from the upper half; side branches 2-8, erectly spreading, each bearing 15-50 flowers; bracts 10-12 mm long, 4 mm broad at the base; floral bracts 6-7 mm long, 2 mm broad at the base; pedicel 7-8 mm long. *Perianth* 18-25 mm long, stipitate for 2 mm, gasteriform basally for more than half its length, thence constricted to a tube of 4-4,5 mm in diameter; gasteriform portion 6-7 mm in diameter, narrow to globose-elliptic, reddish pink, tube light to dark pink; apices white, rarely pink, erectly spreading, obtuse, with green median stripes; inner segments channelled for the length of the gasteriform portion. *Stamens* 16-22 mm long; anthers 2 mm long, included. *Ovary* 5-8 mm long, 2-2,5 mm in diameter, style included; stigma minute, becoming capitate after anthesis. *Capsule* 12-27 mm long, 7 mm in diameter. *Seeds* oblong, 2-5 mm long, 2-3 mm wide. *Flowering time* July to November, but with a peak in September and October.

DISTRIBUTION AND HABITAT

Gasteria pulchra has been recorded in the Gamtoos River valley and adjacent regions of the eastern Cape. The plants are confined mainly to soils derived from Mesozoic conglomerates of the Uitenhage Group (Enon Formation). The main distribution area has hills and rocky outcrops densely vegetated with valley bushveld. *G. pulchra* grows in shady positions on well-drained, acid soils rich in humus. The rainfall, 200-400 mm annually, occurs during both summer and winter. The climate is very hot in summer, but mild and frost-free in winter. Plants observed in their natural habitat near Hankey were scattered on a hilltop in the shade of dense, short, thorny scrub, 1-3 m tall.

DIAGNOSTIC FEATURES AND VARIATIONS

Gasteria pulchra is variable, especially in leaf size, colour and shape. It is characterized by its acaulescent rosette of distinctly channelled, erect to erectly spreading, linear-triangular leaves, obscurely triangular in cross section. The species is further distinguished by its lax panicles with erectly spreading side branches and its reddish pink perianth that is 18-22 mm long. It may be confused with narrow-leaved forms of *G. acinacifolia*, but the latter is a coastal species, much larger in size and with a differently shaped perianth. In the past the name *G. acinacifolia* was often wrongly applied to *G. pulchra*. As with other members of the section *Multifariae*, the juvenile leaves are tuberculate and this characteristic may rarely persist to the adult stage.

G. pulchra is related to *G. excelsa* and *G. nitida*, but the first of these differs in that it has larger triangular-lanceolate leaves (margin sharply serrulate) and its flowering time is December to January. *G. nitida* differs by its short, triangular to triangular-lanceolate leaves (margin entire) and bright reddish pink perianth which is yellowish at the apex. The habitats of the two species are very different and the plants do not occur together. *G. pulchra* has also been confused with narrow-leaved forms of *G. bicolor*, but the latter is distinguished by its foliate stem and smooth juvenile leaves.

CULTIVAR

A xeromorphic form of *Gasteria pulchra* with shorter and more succulent subulate leaves is known (*Rossouw 479*). This form has a simple raceme with bright, reddish pink flowers and has been given the cultivar name 'Studtis'.

VOUCHERS

Hardy 2167 (NBG); *Kramer 361* (PRE); *Rossouw 479* (NBG); *Van Jaarsveld 6193* (NBG); *Wisura 1849* (NBG).

Gasteria pulchra (all parts life size): 1. mature plant; 2. part of inflorescence; 3. capsules and seeds; 4. juvenile plant.

The grassy fynbos habitat of *Gasteria nitida* var. *nitida* near Hankey, eastern Cape.

6. **Gasteria nitida** *(Salm-Dyck) Haw.*: 359 (1827); Bak.: 195 (1880); Bak.: 299 (1896); Berger: 152 (1908); V. Poelln.: 243 (1938a); Jacobsen: 640 (1954); E.J. v. Jaarsveld: 11 (1992). Neo-iconotype: t. 17 of Salm-Dyck, *Monogr. gen. Al.* 29 (1836-1863).

HISTORICAL BACKGROUND

Gasteria nitida was first collected at the Cape by Hendrik Swellengrebel, second son of the Hendrik Swellengrebel who was Governor there from 1739 to 1751. Although born at the Cape, Swellengrebel settled in Doorn, in the province of Utrecht, The Netherlands (Gunn & Codd 1981). He had a keen interest in South African plants, which he grew on his estate Schoonoord. He visited the Cape from 1776 to 1777 and collected the plants, presumably in the Langkloof, in 1776 on his second expedition to the eastern Cape. In 1790 he donated seeds of this species to the Botanic Gardens of Utrecht (Berger 1908).

In 1817 Salm-Dyck described *Aloe nitida* from live plants in his succulent collection, but his description was so generalized and lacking in diagnostic detail that it might have referred to a number of *Gasteria* species. In 1821 he subsequently revised and enlarged his description of *Aloe nitida* (*Obs. bot.*) and again in 1842 (see Stearn 1938; Stafleu & Cowan 1979) in his *Monogr. gen. Al.* (cited 1836-63). This last description was considerably more detailed and accompanied by an excellent colour plate. No type material for any of these descriptions was preserved and this plate was therefore designated as neotype.

DESCRIPTION

Plants acaulescent, decumbent to erect, 60-200 mm tall, 50-280 mm in diameter, solitary or proliferating from the base to form small groups. *Roots* succulent, up to 12 mm in diameter; linear-obclavate, gradually dilating from the tip upwards. *Leaves* at first distichous, becoming rosulate, or remaining distichous, 16-180 mm long, 25-80 mm broad at the base, triangular-lanceolate, rarely lanceolate-acuminate, erectly spreading; adaxial surface canaliculate; abaxial surface somewhat convex with a distinct excentric keel; both surfaces dark green, with faint to dense white spots arranged in irregular transverse bands; epidermis smooth and shiny; margin entire or indistinctly tuberculate; apex acute, mucronate. Juvenile leaves distichous, lorate, 20-40 mm long, 20-35 mm broad at base, erectly spreading at first, becoming patent or recurved; epidermis tuberculate; tubercles domed or subconical, sometimes confluent forming ridges, rarely smooth, dark green, not spotted or rarely spotted; lower surface asperulous; apex obtuse or truncate, mucronate; plants rarely remain in the distichous phase. *Inflorescence* an erectly spreading lax panicle, 0,2-1,2 m long, up to 0,5 m in diameter, or racemose; scape 3-10 mm in diameter at base, branched from the upper half, bearing 1 to 3 erectly spreading side branches; lower bracts 20 mm long, 10 mm broad at the base; floral bracts 8 mm long and 3 mm broad at the base; pedicel 7-8 mm long. *Perianth* 20-25 mm long, bright reddish pink, stipitate for 2-3 mm, gasteriform basally for slightly more than half its length; the gasteriform portion 5-8 mm in diameter (narrow-elliptic), thence constricted into a tube 4-5 mm wide; apices erectly spreading, obtuse, yellowish; inner segments canaliculate for the length of the gasteriform portion. *Stamens* 18-22 mm long; anthers 2 mm long, oblong, not protruding. *Ovary* 6 mm long and 2 mm in diameter; style 13-15 mm long, exceeding the anthers after anthesis; stigma minute, becoming capitate during anthesis, included. *Capsule* 24-30 mm long, 8 mm in diameter. *Seeds* 3-4 mm long, 2 mm wide. *Flowering time* December to February.

DISTRIBUTION AND HABITAT

Gasteria nitida is widely distributed in the south-eastern Cape coastal regions, from near Uniondale in the west to near the Great Fish River mouth in the east, up to an altitude of about 1 000 m. It is often associated with mineral-poor, slightly acid, quartzitic sandstone soils. The plants occur singly or in small groups, in sun or light shade in grassy fynbos, grassveld or renosterveld and on hilly, rocky terrain. When not in flower they are difficult to detect. This species and *G. vlokii* are the only *Gasteria* species confined to an open habitat. The climate is hot in summer and mild in winter. Frost, if it occurs, is slight. The annual rainfall of 600 to 800 mm occurs in summer and winter.

DIAGNOSTIC FEATURES AND VARIATIONS

Gasteria nitida is immediately recognized by its striking, erect to erectly spreading, bright reddish pink racemes. The perianth is yellowish at the throat, a distinct feature not found in any other species of this genus. It is further characterized by its acaulescent rosette of short triangular to triangular-lanceolate, distinctly keeled, erect to erectly spreading leaves. The plants may be solitary or proliferate to form small groups. The inflorescence may be simple or with ascending side branches. The vegetative parts have a chameleon nature (phenotypic plasticity), depending on the prevailing environmental conditions.

When frequent veld fires occur, plants are smaller and tend to remain in the juvenile phase. The leaves are lorate, densely tuberculate and lie flat on the ground, and the plants are very difficult to detect. Without annual fires the leaves grow into a rosette. This is, however, not always the case and a variation exists where the leaves remain distichous (see var. *armstrongii*). This variation is continuous, however, and it grades into the typical *G. nitida*.

G. nitida is closely related to *G. ellaphieae*, *G. carinata* and *G. vlokii*, but these differ by having simple racemose inflorescences, the perianth not being yellow at the apex, and the epidermal texture of the juvenile and adult leaves remaining the same. The leaf is asperulous in *G. vlokii*, tuberculate in *G. ellaphieae* and tuberculate or smooth in *G. carinata*. In floral and vegetative features *G. nitida* comes close to *G. ellaphieae* and *G. vlokii*. These three species flower during December and January and have similar bright red-pink inflorescences. (*G. carinata* flowers during spring.) The distribution of these four related species is clearly allopatric.

G. nitida is also related to *G. pulchra* and *G. excelsa*, having similar juvenile leaves, but the last-named differs in having leaves with very sharp, serrulate leaf margins and large panicles with light pink flowers. *G. excelsa* occurs in shady, forested river valleys. The flowering times of *G. pulchra* and *G. excelsa* coincide and where they grow close together, natural hybrids have been recorded (*Van Jaarsveld 8653*). *G. pulchra* has linear-triangular leaves.

Gasteria nitida var. *nitida*: 1. plant and inflorescence (reduced); 2. longitudinal view of perianth; 3. androecium; 4. gynoecium; 5 & 6. buds in different stages of development; 7. mature perianth; 8. juvenile plant; 9. mature leaf; 10. cross section through leaf; 11. capsule and seed.

Gasteria nitida var. *nitida* prefers an open habitat, generally growing in full sun on grassy or rocky terrain.

Gasteria nitida var. *armstrongii* is well camouflaged and difficult to detect against the stony background it favours.

6a. **Gasteria nitida** var. **nitida**

DESCRIPTION

Leaves rosulate, triangular to triangular-lanceolate with a distinct excentric keel (margin entire). *Inflorescence* racemose, simple or paniculate and up to 1,2 m tall.

DISTRIBUTION AND HABITAT (see also under the species)

Gasteria nitida var. *nitida* in its natural habitat near Loerie can be found in a mosaic of fynbos and renosterveld. The plant community, occurring on an open hillside with a northerly aspect, consists of shrubs 1-1,5 m tall. *G. nitida* var. *nitida* was observed in full sun as well as in light shade.

DIAGNOSTIC FEATURES

This variety is characterized by its acaulescent rosette of triangular to triangular-lanceolate leaves with entire margins.

VOUCHERS

Drège 399 (GRA); *Dyer 2272* (GRA, K, PRE); *Fourcade 2489* (STE); *Van Jaarsveld 7041* (NBG); *Vlok 1836* (NBG).

6b. **Gasteria nitida** var. **armstrongii** *(Schönl.) E.J. v. Jaarsveld*: 12 (1992). *G. armstrongii* Schönl.: 258 (1912). Type: ex hort., from I.L. Drège's garden in Port Elizabeth (K, holo.!).

HISTORICAL BACKGROUND

The variety *armstrongii* was first collected by W. Armstrong in the Humansdorp district in about 1913. It flowered in I.L. Drège's garden in Port Elizabeth in June 1914, from where it was described.

DESCRIPTION

Leaves distichous, patent, lorate; cuticle tuberculate, rarely smooth; apex somewhat retuse, obtuse or truncate. *Inflorescence* racemose, 400-500 mm tall. *Perianth* 20 mm long. *Flowering time* December to January.

DISTRIBUTION AND HABITAT

Gasteria nitida var. *armstrongii* is confined to the coastal flats of the Gamtoos, Jeffrey's Bay and Humansdorp regions, occurring on exposed, stony ground. Near Jeffrey's Bay it can be found on top of a rocky hillside in soils derived from quartzitic sandstone. The vegetation is a combination of fynbos and renosterveld, with karoo elements. The soils are sandy, acid and well drained. The plants grow on stony ground and merge effectively with the background.

At another locality west of the Gamtoos River, plants were observed in alluvial soil. Here they occur in a flat area among round pebbles in grassy fynbos and renosterveld. The plants are often almost geophytic, with only the upper half of the leaves exposed. They grow in full sun to light shade.

DIAGNOSTIC FEATURES

Gasteria nitida var. *armstrongii* is characterized by the distichous, patent to recurved, dark, tubercled leaves that lie flat on the ground, with the leaf apices often retuse. Occasionally plants with smooth and more pointed leaves are encountered. When not in flower these plants may be confused with *G. baylissiana*, but the latter has very different flowers and leaves with dense, white, truncate tubercles.

VOUCHERS

Britten 399/32 (BOL, GRA, K, NBG); *Hall 1530* (NBG); *Moss 19088* (J); *Van Jaarsveld & Marthinus 6960* (NBG); *Van Jaarsveld & Sardien 10977* (NBG).

Gasteria nitida var. *armstrongii* (all parts life size, except 3 and 4): 1. mature plant; 2. inflorescence; 3. perianth segments (x2); 4. androecium and gynoecium (x2).

This steep rock face overlooking the Kouga Dam, in the south-eastern Cape, is the habitat of *Gasteria ellaphieae*.

7. **Gasteria ellaphieae** *E.J. v. Jaarsveld*: 3 (1991b); E.J. v. Jaarsveld: 12 (1992). Type: Cape Province, 20 km west of Patensie, *Van Jaarsveld, Campher & Zeydak 9904* (NBG, holo.!).

HISTORICAL BACKGROUND

Gasteria ellaphieae was discovered by the author in 1989, while he was searching for other plants that are known to occur in the Kouga Dam area of the south-eastern Cape Province.

DESCRIPTION

Plants acaulescent, decumbent to erect, 15-30(-40) mm tall, (50-)70-140 (-160) mm in diameter, solitary or proliferating from the base to form small groups. *Roots* succulent, up to 5 mm in diameter, linear-obclavate, gradually dilating from the tip upwards. *Leaves* at first distichous, becoming rosulate, 20-40(-50) mm long, 10-20 mm broad at the base, firm, triangular to triangular-lanceolate and falcate, the inner erectly spreading, the outer recurved; adaxial surface canaliculate and flat in upper third; abaxial surface somewhat convex with a distinct excentric keel; both surfaces dark green, with dense, white tubercles arranged in irregular transverse bands; epidermis tuberculate; tubercles domed; margin tuberculate-denticulate; apex acute or acuminate, recurved, mucronate. Juvenile leaves distichous, lorate, ascending at first, becoming patent or recurved; tubercles domed or subconical. *Inflorescence* an erectly spreading raceme, 250-400 mm long, simple or branched in upper third and bearing 8-15 flowers; scape 2-2,5 mm in diameter at base; bracts 5 mm long, 2,5 mm broad at the base; floral bracts the same; pedicel 6 mm long. *Perianth* (22-)25-27 mm long, reddish pink, stipitate for 2,5 mm, gasteriform basally for slightly more than half its length; the gasteriform portion 7,5 mm in diameter (narrow-elliptic), thence constricted into a tube 4,5 mm wide; apices erectly spreading, obtuse, white with centric green striations; inner segments canaliculate for the length of the gasteriform portion. *Stamens* 20-25 mm long, three lengthening in advance; anthers 2 mm long, oblong, included or protruding up to 2 mm. *Ovary* 5 mm long and 2 mm in diameter; style 16 mm long, exceeding the anthers after anthesis; stigma minute, curved upwards, becoming capitate during anthesis, included. *Capsule* 12-15 mm long, 5-6 mm in diameter. *Seeds* 3 mm long, 2 mm wide. *Flowering time* January to February.

DISTRIBUTION AND HABITAT

Gasteria ellaphieae is known from the Kouga Dam region in the south-eastern Cape. The terrain is rugged and inhospitable, and plants occur on steep to sheer, shady, rocky south-eastern slopes in mineral-poor, slightly acid, quartzitic sandstone soils. They grow singly or in small groups in light shade, and when not in flower they are difficult to detect among the lichen-covered rocks and leaf debris. *G. ellaphieae* occurs in the centre of species diversity of the genus, and its distribution is sympatric with *G. glomerata*. The latter occurs on vertical cliff faces.

The habitat of *G. ellaphieae* consists of valley bushveld. The plant community includes sclerophyllous shrubs, small trees and herbs. The climate is hot in summer and mild in winter. The annual rainfall of 300-400 mm occurs in summer and winter.

DIAGNOSTIC FEATURES AND VARIATIONS

Gasteria ellaphieae is immediately recognized by its densely tuberculate, triangular, ascending to recurved leaves and erect to erectly spreading, reddish pink racemes, often branched in the upper half. It is an attractive dwarf species, and with its short triangular recurved leaves it sometimes superficially resembles a haworthia. The plants may be solitary or proliferate to form small groups.

It differs from the species with tuberculate adult leaves (*G. carinata*, *G. baylissiana*, *G. nitida* var. *armstrongii* and *G. batesiana*) by its short, triangular and recurved, often falcate leaves. In floral and vegetative characteristics it is similar to *G. nitida* and *G. vlokii*. It has, however, a different leaf shape and epidermal texture (apart from being smaller). The perianth length of *G. ellaphieae* (25-27 mm long) is also somewhat intermediate between the two, which is 29-33 mm in *G. vlokii* and 20-27 mm in *G. nitida*. These three species are immediately distinguished by their bright, red-pink flowers which appear during January and February. *G. nitida* occurs in grassy fynbos regions to the south; it is a much larger species with erect to ascending smooth leaves. *G. vlokii* has asperulous leaves lacking tubercles and occurs in fynbos on the Swartberg mountain range to the west. The distribution of these three related species is clearly allopatric but is confined to rocky, quartzitic sandstone soils.

G. ellaphieae also occurs close to, and is related to, *G. pulchra*. The latter has similar juvenile leaves but differs by being much larger, with ascending, linear, acuminate leaves and a large paniculate inflorescence (flowering in spring).

Superficially, *G. ellaphieae* may also be confused with the larger *G. carinata* and *G. batesiana*, from the renosterveld regions of the southern Cape and from northern Natal and the eastern Transvaal respectively. However, *G. carinata* has a different leaf epidermal texture and the plants flower in spring. *G. batesiana* also has somewhat similar but larger recurved leaves, but here the resemblance ends, as it has a very different leaf epidermal texture and larger flowers.

G. ellaphieae is easily propagated by leaf cuttings grown in sandy, compost-rich soil in light shade. The plant thrives as a pot plant and is a slow-growing but rewarding species.

VOUCHERS

Van Jaarsveld & Sardien 11032, 11084 (NBG).

Gasteria ellaphieae (all parts life size, except 4): 1. inflorescences; 2. mature plant; 3. capsules and seeds; 4. seed (x3); 5. plant in habitat.

The habitat of *Gasteria vlokii* at the top of the Great Swartberg. The plants grow in rock crevices among fynbos vegetation.

8. **Gasteria vlokii** *E.J. v. Jaarsveld*: 170 (1987a); E.J. v. Jaarsveld: 12 (1992). Type: Cape Province, upper slope of Waboomsberg, *Vlok 880* (NBG, holo.!).

HISTORICAL BACKGROUND

Gasteria vlokii was discovered by Jan Vlok, a senior forestry research technician at Saasveld, on 26 December 1984.

DESCRIPTION

Plants acaulescent, proliferating from the base to form small groups up to 140 mm in diameter. *Roots* succulent, up to 8 mm in diameter; linear-obclavate, gradually dilating from the tip upwards. *Leaves* distichous at first, ultimately becoming rosulate, 50-90 mm long, 20-30 mm broad at the base, patent, lorate, lanceolate to triangular, falcate, with an excentric marginiform keel in the upper third, green with dense white spots in obscure transverse bands; adaxial surface flat to slightly convex, becoming channelled during the dry season; abaxial surface convex; epidermis asperulous; margin acute-tuberculate, crenulate, becoming continuous towards apex; apex acute or obtuse with an excentric mucro. Juvenile leaves lorate, apex obtuse, mucronate. *Inflorescence* racemose, curved and spreading, 300-480(-840) mm long, rarely with a pair of side branches; scape biconvex at the base, 4-5 mm in diameter, terete upwards. *Flowers* pendulous, laxly and secundly arranged in the upper half (20 to 36); bracts scarious, lanceolate, acuminate, lower bracts 4 x 7 mm; floral bracts 5 mm long; pedicels 7-10 mm long, pendulous, terete, becoming erect after anthesis. *Perianth* 29-33 mm long, stipitate for 2-3 mm, gasteriform basally for slightly more than half its length, thence constricted to a tube 4 mm in diameter; the gasteriform portion 6-7 mm in diameter, dark reddish pink; apices white with green median striations; apices of inner segments erect, acute, becoming slightly spreading; apices of the outer segments erect. *Stamens* 15-20 mm long; filaments filiform, slightly flattened; anthers 2 mm long, oblong; pollen yellow. *Ovary* 6-7 mm long, oblong, ovoid, six-grooved; style 18 mm long, terete; stigma minute. *Capsule* erect, 15-18 mm long, 6 mm in diameter. *Seeds* oblong, 3-5 mm long, 2 mm wide. *Flowering time* January to February.

DISTRIBUTION AND HABITAT

Gasteria vlokii is known only from the northern slopes of the Great Swartberg and Witteberg. The Great Swartberg is situated in the southern Cape and averages about 1 300-1 800 m in altitude. It consists of a hard quartzitic sandstone and its upper slopes carry mesic mountain fynbos. The vegetation of the lower slopes comprises succulent karoo with a preponderance of *Portulacaria afra*. The plants can be found on both lower and higher slopes and occur singly or in small groups, wedged in crevices and shallow rock pockets in sandy, humic soils with a low mineral content, in shady or exposed situations. When exposed, the leaves become tinged with red. On higher slopes *G. vlokii* is often found growing in moss among tufts of perennial grasses and restiads. The climate here (on the higher slopes) is cool in winter with occasional snow; the summers are hot, but with cool nights. Rainfall, averaging about 400-800 mm per annum, occurs during winter and summer, although summers tend to be drier.

Vegetation in the area consists of evergreen sclerophyll shrubland and the families Restionaceae, Proteaceae and Ericaceae are well represented in the habitat. The area is subject to burning, especially on the higher densely vegetated slopes, but *G. vlokii* growing on the rock faces is well protected, while those plants in less protected areas proliferate from the base after fire. This is the first *Gasteria* species to be found in a mesic mountain fynbos habitat. In spite of the high annual rainfall on the higher slopes, the sheer rock slopes with shallow pockets dry out rapidly, ensuring a semi-arid habitat.

DIAGNOSTIC FEATURES AND VARIATIONS

Gasteria vlokii can be confused with the very variable *G. carinata*. It differs from that species by its patent falcate leaves with asperulous epidermis, its flowering time (January and February), as well as its bright, reddish pink, larger perianth (29-33 mm long). *G. carinata* has erectly spreading, smooth or tuberculate leaves which are not falcate, it flowers in spring, the perianth measures 18-25 mm in length, and the two species are geographically isolated. The asperulous leaf surface of *G. vlokii* is similar to that of *G. disticha*, but here the resemblance ends, as *G. disticha* has distichous leaves and a shorter perianth. *G. vlokii* is also related to *G. nitida* and *G. ellaphieae* (sister species); these flower in December and January, have a bright reddish pink perianth and linear-obclavate, succulent roots. For differences between these three species, see *G. nitida*.

One clone of *G. vlokii* cultivated under controlled greenhouse conditions proved to be self-fertile. Seeds sown in autumn 1986 germinated within three to six weeks.

VOUCHERS

Van Jaarsveld, Vlok & Bayer 8888 (NBG); *Vlok 733* (NBG); *Vlok 880* (NBG); *Vlok 1555* (NBG).

Gasteria vlokii (all parts life size): 1. mature plant; 2. inflorescences; 3. mature leaf; 4. cross section through leaf; 5. capsules and seeds.

The habitat of *Gasteria carinata* var. *carinata*, south-east of Herbertsdale in the southern Cape. *Aloe ferox* (foreground) is a common species of the area.

9. **Gasteria carinata** *(Mill.) Duval*: 6 (1809); Haw.: 87 (1812); Haw.: 203 (1819); Haw.: 357 (1827); Bak.: 192 (1880); Bak.: 296 (1896); Berger: 149 (1908); Pole Evans: t. 291 (1928); V. Poelln.: 237 (1938a); Jacobsen: 629 (1954); Barkhuizen: 187 (1978): Wijnands: 131 (1983). Type: t. 9 of Commelin, *Hort. med. amst.* (1701) designated as lectotype by Wijnands: 132 (1983).

HISTORICAL BACKGROUND

Gasteria carinata var. *carinata* and var. *verrucosa*, with *G. disticha*, were the earliest taxa to have reached Europe. *G. carinata* var. *verrucosa* was first described by Oldenland under the phrase name *Aloe africana flore rubro folio triangulari et verrucosis ab utraque parte albicantibus notato* in his 'kruid boek', a herbarium of South African plants compiled at the Cape probably between 1688 and 1697. Commelin used Oldenland's phrase name in *Horti medici amstelodamensis* (published in 1701), and the description is accompanied by a fine illustration by Jan Moninckx. According to Commelin, the plant was cultivated in the Amsterdam Medicyn-Hoff from seed received from Africa. It is probable that the seed was collected south of the Langeberg by Oldenland when he accompanied Schrijver on his expedition in 1689.

DESCRIPTION

Plants acaulescent, decumbent to erect, 30-180 mm tall, proliferating from the base to form small dense groups 150-600(-800) mm in diameter. *Roots* succulent, up to 6 mm in diameter. *Leaves* distichous at first, becoming rosulate, spirally arranged or remaining distichous, 30-180 (-120) mm long and 10-45(-50) mm broad at the base, triangular to triangular-lanceolate, rarely lorate or ensiform, erectly spreading, lower leaves becoming spreading; adaxial surface plane or channelled; abaxial surface keeled or convex; both surfaces tuberculate or smooth, rarely asperulous, spotted with raised or immersed white, domed tubercles in obscure transverse bands; margin cartilaginous, tuberculate-crenulate, rarely denticulate; apex acute, rarely obtuse, truncate or retuse, mucronate. Juvenile leaves distichous, erectly spreading, smooth or tuberculate, lorate; apex obtuse, mucronate. *Inflorescence* racemose 150-700(-900) mm tall, spreading and slightly curved; scape 4-12 mm broad at the base; bracts 6-15 mm long, 3-5 mm broad at the base; floral bracts 6-12 mm long, 2,4 mm broad at the base, piliferous; pedicel 7-15 mm long. *Perianth* variable, 16-27 mm long, stipitate for 2-3 mm, gasteriform basally for more than half its length, thence constricted to a tube 3-5 mm in diameter; the gasteriform portion pink, narrow-elliptic, rarely globose-elliptic, 6-8 mm in diameter; the tube white with green striations; apices of inner and outer segments obtuse, erectly spreading, light pink or white with centric green striations; margin of inner segments free and channelled for the length of the gasteriform portion. *Stamens* oblong, 2 mm long, and included. *Ovary* 6-7 mm long and 2.5 mm broad; style 14 mm long. *Capsule* 19-23 mm long, oblong, and 7 mm broad. *Seeds* 3-4 mm long and 2 mm wide. *Flowering time* July to November.

DISTRIBUTION AND HABITAT

Gasteria carinata is widely distributed in the southern Cape south of the Riviersonderend and Langeberg mountains. The species is found from close to sea-level up to approximately 300 m in altitude. The habitat consists of hilly terrain; the natural vegetation is renosterveld and occasionally limestone fynbos. *G. carinata* occurs in dry river valleys, on shale banks and outcrops in the shade of xerophytic shrubs, together with other succulent species. The rock formations consist of Palaeozoic shale and Cenozoic limestone. It is also found to a lesser extent on Mesozoic conglomerates. Rainfall, between 300 and 500 mm per annum, occurs in winter and summer, but the summers tend to be dry. Winters are moderate, and if frost occurs, it is slight. Summers are hot, with cool nights. The plants, which grow in small dense groups almost always in the protection of herbs and shrubs, are well camouflaged and difficult to locate. Occasionally plants may be found in the open.

DIAGNOSTIC FEATURES AND VARIATIONS

Gasteria carinata is the most variable species in the genus, and it is difficult to define clearly as populations vary continuously. Except for the variety *retusa*, however, it is the only species that occurs south of the Langeberg and therefore cannot be confused with any other. *G. carinata* can be recognized by the following characteristics: plants acaulescent with triangular to triangular-lanceolate, tuberculate or smooth (rarely asperulous) leaves which are rosulate or spirally arranged (rarely distichous). Juvenile and adult leaves always have a similar texture. The plants are prolific from the base to form small groups. In floral characters the species is also variable, especially in size. It closely resembles *G. nitida*, *G. ellaphieae* and *G. vlokii*, but these flower in December and January (bright red-pink). *G. vlokii* has patent, falcate leaves with an asperulous leaf epidermis. *G. nitida* has a branched inflorescence (perianth with yellowish apices) and juvenile leaves that change from tuberculate to smooth in the adult stage. *G. ellaphieae* has short, triangular, recurved and tuberculate leaves.

G. carinata also resembles *G. disticha* and *G. brachyphylla* (which occur north of the Langeberg), but it can be distinguished from those by its smaller stature and different leaf shape and texture. *G. brachyphylla* and *G. disticha* are larger, usually with broad (40-80 mm), lorate, smooth or asperulous leaves which remain distichous. Three varieties of *G. carinata* are recognized, based on leaf shape, phyllotaxy and epidermis texture. Differences are discussed under each variety.

Gasteria carinata var. *carinata* (all parts life size): 1. mature plant; 2. part of inflorescence; 3. capsules; 4. mature leaf of var. *retusa*; 5. mature leaf of var. *verrucosa*; 6. cross section through mature leaf of var. *carinata*; 7. cross section through 4; 8. cross section through 5.

A large colony of *Gasteria carinata* var. *carinata* growing in renosterveld on an embankment of the Great Brak River, southern Cape.

9a. **Gasteria carinata** var. **carinata**

DESCRIPTION

Leaves at first distichous, rarely remaining distichous, becoming rosulate or spirally arranged, (30-)50-180 mm long and 25-100 mm broad at the base, triangular to triangular-lanceolate or lorate, with a more or less distinct keel; adaxial surface channelled; epidermis with raised or immersed white tubercles; apex acute or obtuse.

DISTRIBUTION AND HABITAT

Gasteria carinata var. *carinata* is found south of the Langeberg, in the shade of xerophytic shrubs on shale banks. The typical variety with triangular, tuberculate leaves was observed near Bredasdorp on the rocky shale bank of a dry stream, in open shrubby vegetation about 1 m tall, interspersed with taller shrubs. This variety grows in the shade of these shrubs in humus-rich soil and is often very well camouflaged against the background. Kobus Venter (pers. comm.) observed *G. carinata* growing on a quartz outcrop.

In another part of its range near Mossel Bay, this taxon grows in noorsveld, a dense xerophytic scrub with a preponderance of succulents. Plants were observed on a steep, well-drained eastern slope (Cenozoic alluvium), in the shade of shrubs as well as in the open.

DIAGNOSTIC FEATURES AND VARIATIONS

Gasteria carinata var. *carinata* is variable, grading into the other varieties; intermediates are often encountered. In the typical form the leaves are triangular, tuberculate and keeled. At the western end of its distribution they tend to be more distinctly tuberculate than they are further east.

CULTIVARS

A number of very attractive local forms exist, of which the following have been brought into cultivation:

'Kykoedie' A small form was introduced from Kykoedie by Bruce Bayer. It has short and broad, triangular, tuberculate leaves 60 mm long.

'Gourits' This small form from the Gourits River area has smooth, broad, triangular, lorate leaves 30 mm long and with an obtuse apex.

'Mossel Bay' This beautiful, prolific plant was collected by Professor Ted Schelpe from Mossel Bay in the eastern part of the distribution range. The plants have smooth, linear-lorate leaves, 90-120 x 15-20 mm, with an obtuse leaf apex and dense spotting in attractively arranged transverse bands.

'Great Brak' Originally from the Great Brak River area, a prolific clone was brought into cultivation at Stellenbosch University Botanic Garden. It was later described as *G. schweickerdtiana* V. Poelln. The plants have smooth lorate, very faintly spotted leaves, 70-100 x 20-25 mm, with an obtuse apex.

'Klein Brak' Howard Gie introduced a form from Kleinvlei near Little Brak River. It has smooth, faintly spotted, triangular-lanceolate leaves, 80-110 x 40-45 mm. It occasionally produces offshoots from the base.

VOUCHERS

Bayer 2253 (NBG); *Burgers 2275* (STE); *Kramer 308* (PRE); *Van Jaarsveld 7029* (NBG); *Vlok 602* (NBG).

9b. **Gasteria carinata** var. **retusa** *E.J. v. Jaarsveld*: 15 (1992). Type: Cape, Orange Grove, *Van Jaarsveld & Stayner 4656* (NBG, holo.!).

HISTORICAL BACKGROUND

The variety *retusa* was first collected by Hugo Naudé in August 1924, at Keerom Buttress.

DESCRIPTION

Leaves distichous, 50-90 mm long, 25-35 mm broad at the base, lorate, with raised semi-translucent, white, dome-shaped tubercles; margin rugulose; adaxial surface plane; abaxial surface convex or plane; apex truncate to retuse. *Inflorescence* 300-450 mm long. *Perianth* variable, gasteriform portion narrow to globose-elliptic.

DISTRIBUTION AND HABITAT

Gasteria carinata var. *retusa* occurs in mountainous terrain, on the lower southern slopes of mountains near Worcester and in the Heidelberg district further east. The habitat consists of karroid brokenveld, with a preponderance of succulents. The plants grow on steep shady rocks, and the shale-derived soil is rich in humus.

DIAGNOSTIC FEATURES AND VARIATIONS

This variety is distinguished by its distichous, distinctly tuberculate, lorate leaves, with undulating margins, which are retuse or truncate at the apex. It occurs in close proximity to *Gasteria disticha* and there are similarities in leaf and perianth shape between these two taxa. The tuberculate nature and the smooth (not asperulous) surfaces of the leaves in *G. carinata* var. *retusa* are, however, more typical of *G. carinata*. One gathering is somewhat intermediate between *G. disticha* and var. *retusa*; this may be due to hybridization.

VOUCHERS

Lavis s.n. (BOL); *Stayner 4512* (NBG); *Stayner & Littlewood 427/60* (NBG); *Van Breda 78* (PRE); *Van Jaarsveld & Stayner 4656* (NBG).

9c. **Gasteria carinata** *(Mill.) Duval* var. **verrucosa** *(Mill.) E.J. v. Jaarsveld*: 15 (1992). *Gasteria verrucosa* (Mill.) Duval (1809). *Aloe verrucosa* Mill.: 20 (1768). *Aloe linguiformis* DC. var. *verrucosa* (Mill.) DC.: 68 (1799). Type: t. 2 of p. 131 (no. 36) in Boerhaave, *Index alter plantarum* (1720).

HISTORICAL BACKGROUND

Gasteria carinata var. *verrucosa* was first described by Oldenland in his 'kruid boek' under the phrase name *Aloe africana glabro folia minutissimus cavitatibus donato* (Valentyn 1726). It was probably collected by Oldenland during Schrijver's expedition in 1689. It is very likely that it

Orange Grove, north of Worcester, where *Gasteria carinata* var. *retusa* may be found growing among other succulent species.

The typical habitat of *Gasteria carinata* var. *verrucosa* in the southern Cape. The plants occur in dry fynbos or in a mosaic of subtropical thicket and fynbos.

was introduced to the Amsterdam Medicyn-Hoff in the late seventeenth century, as two very fine illustrations of this species appear in the Moninckx Atlas, 6: t. 12 and 7: t. 48 (1686-1706), one being depicted by Maria Moninckx (Wijnands: 132, 1983).

DESCRIPTION

Leaves distichous, rarely becoming spirally arranged, 30-70(-180) mm long, 15-30(-35) mm broad at base, linear-lanceolate to lorate, distinctly tuberculate due to raised, semi-translucent, white or green, domed tubercles; epidermis green to glaucous-green, smooth or asperulous; adaxial surface plane to channelled; abaxial surface without keel. *Inflorescence* 120-300 mm long. *Perianth* 20-25 mm long. For other characteristics, see var. *carinata*.

DISTRIBUTION AND HABITAT

Gasteria carinata var. *verrucosa* occurs on soils derived from shales, limestone, and conglomerate of the Bredasdorp Formation. It is found on rocky outcrops as well as in flat areas in renosterveld, limestone fynbos and outliers of subtropical thicket. Plants were observed occurring in light shade in renosterveld. The soil is rich in humus and well drained.

In one locality near the mouth of the Breede River *G. carinata* var. *verrucosa* occurs commonly in isolated patches of subtropical thicket. It grows in small groups in shade and rarely also in sun. The shale soils are well drained and rich in humus.

In another locality near De Hoop *G. carinata* var. *verrucosa* was observed on flat, rocky (calcrete) terrain in a transition zone of renosterveld and limestone fynbos. The vegetation consists of an open shrubby layer, 1-3 m tall, interspersed with smaller plants. The variety *verrucosa* is found in small to large dense colonies in the shade of sclerophyllous shrubs. The soil is sandy, humus-rich and alkaline.

DIAGNOSTIC FEATURES AND VARIATIONS

Gasteria carinata var. *verrucosa* is distinguished by its linear-lanceolate leaves which are distinctly tuberculate, and often asperulous; they are normally distichous, rarely becoming spirally arranged. The plants are very variable in size and leaf shape, so this variety grades into the others; it is recognized therefore only at varietal rank.

There is a gradual change in leaf colour and increase in tubercle size in plants found along the Breede River from Malgas to its mouth. At Malgas the tubercles are small and greenish but towards the river mouth the plants become more glaucous and very densely tuberculate, reminiscent of the form commonly found in cultivation.

CULTIVARS

With its densely tubercled, distichous leaves, *G. carinata* var. *verrucosa* is one of the most attractive species in cultivation. Some of the local variations and forms include:

'Malgas' Plants from Malgas brought into cultivation at Kirstenbosch have lorate leaves up to 40 mm long.

'Infanta' A very large form from Infanta, with leaves up to 280 mm long and 35 mm broad at the base and densely covered with white tubercles, has been brought into cultivation.

'Port Beaufort' A small form from the eastern bank of the Breede River mouth (Port Beaufort), with lorate leaves up to 30 mm long, was brought into cultivation at Kirstenbosch. It has greenish white tubercles and an obtuse leaf apex.

'Albertinia' Bruce Bayer collected plants from Albertinia for cultivation; they have lorate triangular leaves with acute apices, 60-90 mm in length and 25 mm broad at the base, and greenish white tubercles in obscure transverse bands.

'Herbertsdale' In the vicinity of Herbertsdale, plants have erectly spreading, linear, acuminate, deeply channelled leaves, 80-140 mm long, with dense, green and whitish tubercles in obscure transverse bands. The leaf margin is truncate (double-keeled). This form corresponds very well with Thunberg's specimen (UPS) described by N.E. Brown as *Gasteria thunbergii*.

VOUCHERS

Bayer 2673 (NBG); *Burgers 2294* (STE); *Marsh 817* (PRE, STE); *Thunberg 8589* (UPS); *Van Jaarsveld 7559* (NBG).

The top of the Numeesberg, in the Richtersveld, is the habitat of *Gasteria pillansii* var. *pillansii*.

10. **Gasteria pillansii** *Kensit*, in Bolus & Kensit: 163 (1909); V. Poelln.: 244 (1938a); Jacobsen: 641 (1954); Court: 177 (1981); E.J. v. Jaarsveld: 17 (1992). Type: Cape Province, Clanwilliam, *Pillans 833* (BOL, holo.!, PRE, iso.).

HISTORICAL BACKGROUND

The first recorded collection of *Gasteria pillansii* was that of N.S. Pillans in the Clanwilliam district in November 1906. A reproduction of an early illustration of *G. pillansii* appeared in 1967 in the *Catalogue of pictures in the Africana Museum*. The illustration was probably the work of Heinrich Claudius, a German naturalist who, in the service of the Dutch East India Company, accompanied Oloff Bergh on his expedition to Namaqualand from 27 August to 24 October 1683.

DESCRIPTION

Plants acaulescent, decumbent to erect, 50-200 mm tall and 60-400 mm in diameter, proliferating from subterraneous stolons and forming dense clumps (comprising 10 to 150 plants per clone) up to 1 m in diameter. *Roots* succulent. *Leaves* distichous, 20-200 mm long and 15-50 mm broad at the base, lorate, erectly spreading or patent; adaxial surface becoming somewhat channelled in the dry season; abaxial surface convex; both surfaces spotted with immersed tubercles in obscure transverse bands; epidermis asperulous, rarely with few domed tubercles; margin cartilaginous, tuberculate or crenulate; apex obtuse to acute, mucronate, rarely truncate or retuse. *Inflorescence* racemose, one to three per plant, spreading and slightly curved, 0,06-1,2(-1,65) m long, rarely with a pair of side branches; bracts 7-9 mm long, 4-6 mm broad at the base. *Flowers* laxly and secundly arranged in upper half (12 to 40 per raceme), pedicel 9-20 mm long; floral bracts 4-10 mm long and 2-4 mm broad at the base. *Perianth* 25-45(-50) mm long and 6-8 mm in diameter, stipitate for 3 mm, obscurely gasteriform basally for a third or less of its length; the gasteriform portion pink, globose (sometimes indistinct), or globose-elliptic and slightly constricted above the ovary; thence gradually enlarging again to the same diameter as gasteriform portion (rarely clavate); tube white, with green striations; apices obtuse, erectly spreading or becoming recurved, light pink with green median stripes, membranous margin of the three inner segments abruptly constricted 4 mm from the base. *Stamens* 22-32 mm long, filaments slightly flattened, becoming contracted after anthesis; anthers orange-yellow, 2-3(-5) mm long, included or exserted up to 5 mm; pollen yellow. *Ovary* 6-8 mm long; style 18-25(-30) mm long; stigma minute. *Capsule* oblong, 15-23 mm long, 7 mm broad. *Seeds* 4-5 mm long, 2,5 mm wide. *Flowering time* November to April.

DISTRIBUTION AND HABITAT

Gasteria pillansii is endemic to the dry Namaqualand region where it is found at elevations from close to sea-level to about 1 000 m on the Namaqualand escarpment. Rainfall, ranging from 25 to 200 mm per annum, occurs from April to September. Heavy fog from the Benguela Current provides extra moisture. Light frost may occur in the area. Summers are very hot, with temperatures often reaching 40 °C.

G. pillansii has been found growing on quartzitic sandstone, granite, and shales. The soils are sandy and well drained. The plants are confined mainly to the cooler southern and eastern aspects and are almost always found beneath xerophytic shrubs or in the shade of rocks, rarely in exposed or north-facing aspects. The vegetation is xerophytic and consists mainly of succulent karoo and Namaqualand brokenveld. *G. pillansii* has also been found in arid fynbos and strandveld.

DIAGNOSTIC FEATURES AND VARIATIONS

Gasteria pillansii is distinguished by its distichous lorate leaves and its large perianth which is 25-45(-50) mm long and indistinctly gasteriform for less than a third of its length. The inner perianth segments are constricted below the apex of the ovary, a diagnostic feature that immediately separates this from any other species. It is further characterized by the asperulous epidermis of its leaves. The plants are stoloniferous and occur in dense, small to large groups. There is a marked vegetative resemblance to *G. disticha* of the southern Great Karoo, which has similar leaves, but a smaller perianth, 12-20 mm long, which is distinctly gasteriform basally for two thirds of its length. In floral characters, *G. pillansii* can be confused with *G. acinacifolia*, *G. croucheri* and *G. batesiana* from the eastern Lowveld regions of South Africa. These species have a somewhat similar large perianth, but *G. pillansii* differs from them in that its perianth is obscurely gasteriform for less than a third of its length. In other respects *G. acinacifolia*, *G. croucheri* and *G. batesiana* are quite different, having rosulate, keeled leaves and a narrowly elliptic perianth, gasteriform for half its length.

G. pillansii is very variable, especially in leaf and perianth size. This variation is continuous and *G. ernesti-ruschii* has been treated here as a variety of *G. pillansii*. Towards the coast, the leaves become more markedly asperulous and covered with sand and dust. One population near Vredendal has leaves with raised white tubercles (*Mitchell 713/84*, NBG). Two varieties are recognized, mainly by quantitative features as well as by flowering time.

10a. **Gasteria pillansii** var. **pillansii**

DESCRIPTION

Leaves 70-200 mm long. *Racemes* 0,3-1,2(-1,65) m long. *Perianth* 30-40(-50) mm long. *Flowering time* November to January.

DISTRIBUTION AND HABITAT

The variety *pillansii* is distributed in a 100- to 150-km belt that lies parallel to the west coast, from the Clanwilliam region in the south to the Richtersveld in the north. Some populations near Clanwilliam occur on flat, rocky, quartzitic sandstone ridges. *G. pillansii* var. *pillansii* grows in sandy humus-rich soil, forming dense clusters of up to 1 m in diameter.

DIAGNOSTIC FEATURES AND VARIATIONS

The variety *pillansii* is very variable in floral and leaf characters.

Gasteria pillansii var. *pillansii* (all parts life size, except 5, 6 and 7): 1. mature plant; 2. raceme; 3. part of infructescence; 4. juvenile plant; 5. perianth segments and stamens (x2); 6. stipe (x2); 7. gynoecium (x2); 8. capsule and seed.

A flowering specimen of *Gasteria pillansii* var. *pillansii* at the top of the Numeesberg, Richtersveld, at the northern end of its range.

A barren-looking landscape near Lorelei, southern Namibia, supports *Gasteria pillansii* var. *ernesti-ruschii* which grows among rocks in the succulent karoo.

CULTIVARS

A number of local forms are encountered.

'Clanwilliam' In the Clanwilliam region, plants attain their largest size with leaves up to 200 mm long and the perianth up to 50 mm long.

'Krakadou' A natural mutation with an amalgamation of the leaf spots was found in a normal population south of Clanwilliam.

'Vredendal' Anthony Mitchell collected an aberrant form with tuberculed leaves near Vredendal.

'Rosyntjiesberg' At the top of the Rosyntjiesberg a form with broad short leaves was collected by the author.

VOUCHERS

Brandham 3123 (NBG); *Compton 18866* (NBG); *Peers 1074/30* (BOL, NBG); *Pillans 833* (BOL, NH, PRE); *Van Jaarsveld 9808* (NBG).

10b. **Gasteria pillansii** var. **ernesti-ruschii** (*Dinter & V. Poelln.) E.J. v. Jaarsveld:* 17 (1992). *Gasteria ernesti-ruschii* Dinter & V. Poelln., in V. Poelln.: 239 (1938a). Iconotype: Namibia, Lorelei, *Ernest Rusch s.n.* Illustration in *Kakteen* 63: 240.

HISTORICAL BACKGROUND

The varietal epithet commemorates Ernest Rusch, who first collected this variety in 1938 in the southern Namib.

DESCRIPTION

Leaves 20-70 mm long. *Racemes* 60-300 mm long. *Perianth* 25-30 mm long. *Stamens* shortly exserted. For other characteristics, see var. *pillansii*. *Flowering time* December to April, peaking in February and March.

DISTRIBUTION AND HABITAT

Gasteria pillansii var. *ernesti-ruschii* is found along the coastal mountains of the Richtersveld, in succulent karoo. A rich succulent flora, possibly the richest in the world, predominates in this area. Some plants occur on steep, rocky, southern and eastern aspects in quartzitic rock crevices.

DIAGNOSTIC FEATURES AND VARIATIONS

The variety *ernesti-ruschii* is distinguished from the typical variety by its dwarf stature, leaves shorter than 70 mm, and its flowering time, which is mainly in February and March. It is also very variable and merges into var. *pillansii*.

CULTIVARS

The following forms have been brought into cultivation:

'Oograbies' Variations (clines) from the coastal Oograbies mountain range in the Port Nolloth district have narrow, lorate leaves 20-40 x 15-20 mm. They flower during March and April.

'Lekkersing' Ben Groen collected an attractive form from Lekkersing in the southern Richtersveld. It has short, broad, truncate leaves with a rugulose margin.

VOUCHERS

Groen & Aleva 166 (NBG); *Herre 5667* (NBG); *Mitchell 915* (NBG); *Oliver, Toelken & Venter 433* (PRE); *Van Jaarsveld, Perry & Snijman 4152* (NBG).

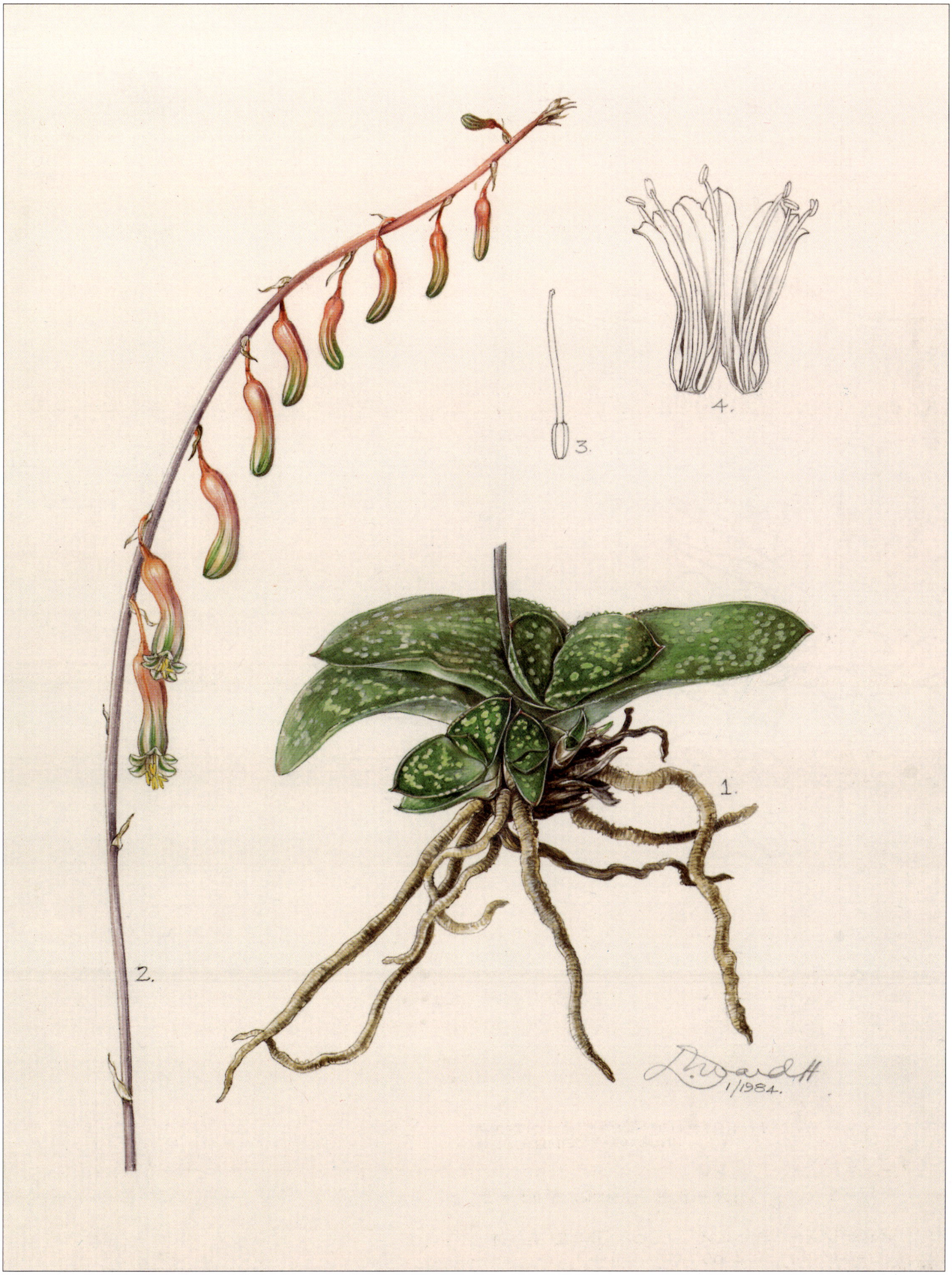

Gasteria pillansii var. *ernesti-ruschii* (all parts life size, except 3 and 4): 1. cluster of plants; 2. inflorescence; 3. gynoecium (x1,5); 4. perianth segments and stamens (x1,5).

Renosterveld such as this at Langvlei, west of Robertson, is typical habitat of *Gasteria disticha*.

11. **Gasteria disticha** *(L.) Haw.*: 352 (1827); Bak.: 186 (1880); Bak.: 291 (1896); V. Poelln.: 239 (1938a); Jacobsen: 631 (1954); Bayer: 30 (1979); Wijnands: 132 (1983). Lectotype: t. 8 of Commelin's *Hort. med. amst.* (1701), fide Wijnands: 131 (1983).

HISTORICAL BACKGROUND

Gasteria disticha was one of the earliest Cape species to reach Europe. It was first described between 1688 and 1697 by Oldenland in his 'kruid boek' under the phrase name *Aloe africana flore rubro folia maculis ab utraque parte albicantibus notato*. The plant was collected by Oldenland when he accompanied the Schrijver expedition in 1689 (Wijnands 1983: 132). Commelin used Oldenland's phrase name in *Horti medici amstelodamensis* 2, published in 1701. The description is accompanied by a fine illustration by Jan Moninckx.

DESCRIPTION

Plants acaulescent, decumbent to erect, 25-180(-230) mm tall and 75-180(-230) mm in diameter, proliferating from the base to form small groups. *Roots* succulent, up to 5 mm in diameter. *Leaves* distichous, 60-120(-170) mm long and 30-45 mm broad at the base, lorate, erectly spreading, rarely somewhat recurved, the apices becoming incurved during the dry season; upper surface convex to flat, becoming channelled during the dry season; lower surface convex or flat; both surfaces with dense white spots arranged in irregular transverse bands; epidermis asperulous; margin irregularly undulating, tuberculate-crenulate; apex obtuse, rarely truncate, mucronate. Juvenile leaves often patent or recurved, asperulous; with a truncate apex. *Inflorescence* racemose, simple, or with a pair of side branches, 200-900 mm tall, bearing 150 to 180 flowers on upper half; floral bracts 7 mm long, 2 mm broad at the base; pedicels 5-7 mm long. *Perianth* 12-20 mm long, stipitate for 2-3 mm, gasteriform basally for more than half its length, thence constricted to a tube of 3-4 mm in diameter; gasteriform portion globose-elliptic or narrow-elliptic, 6 mm in diameter, pink to reddish pink; tube white, with green striations; apices erectly spreading, obtuse; margin of inner segments free, canaliculate for the length of the gasteriform portion. *Anthers* oblong, 2 mm long, not protruding. *Style* 7-10 mm long; stigma minute, included, exceeding the anthers after anthesis. *Capsule* oblong, 15-23 mm long, 7 mm broad. *Seeds* 3-4 mm long, 2-3 mm wide. *Flowering time* July to February, but mainly in September.

DISTRIBUTION AND HABITAT

Gasteria disticha is widely distributed in the southern parts of the Great Karoo, from Beaufort West in the north to the Swartberg Mountains in the south, and again in the Worcester and Robertson districts. It is found at elevations of up to 800 m in karroid brokenveld and renosterveld, occurring in the shade of karoo shrubs on rocky outcrops, on hilly terrain, flat areas, and rock formations of tillite, sandstone, mudstone and shales, as well as on dolomite outcrops. The soils are minerally rich and well drained. Rainfall, ranging from 100 to 300 mm per annum, occurs in winter and summer. In the south-west the summers tend to be dry, but elsewhere the reverse is true. Temperatures are very hot in summer and moderate in winter, but frost may be experienced, especially in the northern parts. Plants from a particular locality near Worcester were observed growing in karroid brokenveld, below short xerophytic shrubs and herbs and in filtered sunlight. In the most northerly distribution area, *G. disticha* grows among dolerite rocks on the southern slope of a koppie. At another locality, plants were found on top of a mountain, in renosterveld.

DIAGNOSTIC FEATURES AND VARIATIONS

Gasteria disticha is characterized by its distichous, lorate, obtuse leaves which have a distinct asperulous surface. The leaves are often undulating, especially the margins, which are irregularly wrinkled. It closely resembles *G. brachyphylla*, but the latter is immediately distinguished by its smooth, usually acute leaves. It is variable in perianth shape, the gasteriform portion sometimes narrow-elliptic (southern forms), and almost indistinguishable from members of the series *Multifariae*. Superficially it resembles *G. pillansii* from Namaqualand, and differences are discussed under that species. *G. disticha* is variable, with several local forms.

CULTIVARS

'Nuy' Plants from Worcester usually have undulating lorate leaves, and flower from August to September.

'Pieter Meintjies' A form with shiny flowers was collected by Bruce Bayer at Pieter Meintjies, west of Matroosfontein.

'Beaufort West' Plants from Beaufort West are more robust, with broad, short, erectly spreading leaves. This form flowers in December.

VOUCHERS

Archer 525 (BOL); *Lavranos & Bleck 20901* (NBG); *Marloth 8935* (NBG, PRE); *Perry 1573* (NBG); *Van Jaarsveld 7041* (NBG).

Gasteria disticha (all parts life size, except 3, 4 and 5): 1. mature plant; 2. inflorescence; 3. perianth segments and stamens (x2); 4. gynoecium (x4); 5. bract (x3); 6. section through leaf; 7. capsule and seeds.

Gasteria brachyphylla var. *bayeri* shares its succulent karoo habitat near Vanwyksdorp with guarri shrubs (*Euclea undulata*) and *Aloe striata*.

12. **Gasteria brachyphylla** *(Salm-Dyck) E.J. v. Jaarsveld*: 19 (1992). *Aloe brachyphylla* Salm-Dyck: 29 t. 8 (1840, see Stafleu & Cowan 1979), cited 1836-1863. Iconotype: t. 8 of Salm-Dyck, *Monogr. gen Al.* (1836-1863).

HISTORICAL BACKGROUND

Gasteria brachyphylla was most probably first collected by Oldenland in the Willowmore district during the Schrijver expedition in 1689 and introduced to the Amsterdam Medicyn-Hoff during the late seventeenth century. It was first cultivated in England by Mr Miller in 1731 (Aiton 1789).

The name of this taxon presented some difficulty at first during the course of research for this revision, since the well-established name *Gasteria nigricans* Haw. could not be upheld. When Haworth (1804) described his *Aloe nigricans* based on Aiton's (1789) *Aloe lingua* var. *crassifolia*, he had in mind *G. brachyphylla*. Unfortunately, Aiton based his var. *crassifolia* on a plate in Commelin's *Horti medici amstelodamensis* 2 (1701), which is also the type of *G. disticha*. *G. nigricans* and *G. disticha* were thus based on the same element and the epithet *nigricans* could not be upheld.

G. brachyphylla (Salm-Dyck) E.J. v. Jaarsveld appeared to be the oldest legitimate name that took priority after *G. nigricans*. Salm-Dyck's accompanying plate (t. 8) of his *Aloe brachyphylla* (*Monogr. gen. Al.*, 1836-63) agrees well with his description and was therefore designated as iconotype. *G. brevifolia* Haw. (1812) and *G. crassifolia* Haw. (1827) could also very well belong to this taxon. Haworth's descriptions upon which these taxa are based are, however, without any type material and are so short, generalized and lacking in diagnostic detail that they might refer to more than one species.

DESCRIPTION

Plants acaulescent, decumbent to erect, 25-230 mm tall, 75-230 mm in diameter, prolific from the base to form small groups. *Roots* succulent, up to 5 mm in diameter. *Leaves* distichous, 15-230 mm long, 22-80 mm broad at the base, lorate, rarely triangular-lanceolate; inner leaves erect to erectly spreading, becoming patent (rarely recurved); apices becoming incurved during the dry season; adaxial surface convex to flat; abaxial surface flat; epidermis smooth, dark green; both surfaces with dense white spots arranged in obscure transverse bands; margin wavy, crenulate, becoming continuous towards the apex; apex acute, obtuse or truncate. Juvenile leaves lorate, asperulous, and densely spotted; apex obtuse. *Inflorescence* racemose, simple or with a pair of side branches, 0,2-0,9(-1,11) m tall; pedicel 5-7 mm long. *Perianth* 12-22 mm long, stipitate for 2-3 mm, gasteriform basally for more than half its length, thence constricted to a tube 3-4 mm in diameter; the gasteriform portion pink, 5-7 mm in diameter, globose or globose-elliptic; tube white with green striations; apices erectly spreading. *Anthers* oblong, 2 mm long, not protruding. *Style* 7-10 mm long, included; stigma minute, exceeding the anthers after anthesis. *Capsule* oblong, 15-23 mm long, 7 mm broad. *Seeds* 3-4 mm long, 2-3 mm wide. *Flowering time* July to February, but mainly in September and October.

DISTRIBUTION AND HABITAT

Gasteria brachyphylla is widely distributed in the Little Karoo, at elevations of up to 1 000 m. It occurs in flat to mountainous terrain, with scattered dwarf xerophytic shrubs interspersed with taller shrubs and small trees in succulent karoo, spekboomveld, karroid brokenveld and renosterveld. It grows mainly on shales and sandstones. It has also been found to a lesser extent on Cenozoic alluvium, Mesozoic conglomerates and Namibian limestones. The soils are sandy and well drained. The climate is hot and dry with mild, almost frost-free winters. Precipitation, ranging between 200 and 300 mm per annum, occurs in summer and winter, but there is a tendency to summer aridity.

DIAGNOSTIC FEATURES AND VARIATIONS

Gasteria brachyphylla is a variable taxon with the following diagnostic characteristics: it is acaulescent with dark green, smooth and shiny, distichous leaves (margin often wavy). The typical variety *brachyphylla* has an acute leaf apex. It is closely related to *G. disticha* and *G. bicolor*, but *G. disticha* has a distinctly asperulous leaf surface and an obtuse apex. *G. bicolor* differs mainly in that it has a leafy stem, and leaves that are often twisted sideways and have a serrulate margin. *G. brachyphylla* may also be confused with *G. carinata*, but the latter usually has spirally arranged, keeled and tuberculate or smooth leaves; when distichous, the leaves are tuberculate. *G. carinata* also differs by having a larger perianth, which has a narrow-elliptic gasteriform portion.

12a. **Gasteria brachyphylla** var. **brachyphylla**

DESCRIPTION

Leaves 90-230 mm long, smooth and shiny; apex acute, rarely obtuse. *Inflorescence* racemose or with a pair of side branches. *Flowering time* July to December, but with a peak in September and October.

DISTRIBUTION AND HABITAT

The variety *brachyphylla* is widely distributed in the Little Karoo and can be found in the shade of karoo shrubs in well-drained soils. South of Willowmore it was found growing in southern Cape renosterveld, on hilly terrain. It occurs among rocks in the shade of xerophytic shrubs, occasionally in full sun. Most of the plants of var. *brachyphylla* found in this area were extensively damaged by porcupines, which uproot them to savour the soft, palatable inner crown. Fortunately new plants proliferate from broken leaf pieces.

DIAGNOSTIC FEATURES AND VARIATIONS

Gasteria brachyphylla var. *brachyphylla* is very variable in leaf size. The plants are readily distinguished from *G. disticha* by their smooth, darker green leaves with acute apices (see under the species).

In the Willowmore and Barrydale districts, plants are more xeromorphic, and less prolific from the base; the leaves are dark green with white, semi-translucent, continuous margins towards the apices. Populations north of Calitzdorp have narrow, almost linear leaves, and

Gasteria brachyphylla var. *brachyphylla* (all parts life size): 1. mature plant; 2. inflorescence; 3. capsules and seeds; 4. apex of leaf; 5. cross section through mature leaf.

Aloe ferox plants dominate typical *Gasteria brachyphylla* var. *brachyphylla* habitat near Ladismith in the Little Karoo. The gasterias usually grow in the shade of xerophytic shrubs.

The obtuse leaves of *Gasteria brachyphylla* var. *bayeri* are a distinguishing feature.

those populations east of the Willowmore district resemble *G. bicolor* from the Jansenville and Steytlerville regions, although the latter has a leafy stem and linear, lorate leaves.

CULTIVARS

A number of local forms exist, of which the following noteworthy plants have been brought into cultivation:

'Brandrivier' An attractive form from the farm Brandrivier in the Barrydale district was introduced into cultivation by Harry Hall. It has short, dark green, acute leaves 120 mm long, and flowers during November.

'Cango' Plants from the district of the Cango Caves and Oudtshoorn have linear, lorate leaves up to 200 mm long, with acute apices, and the inflorescence is often branched. They flower in September.

'Willowmore' Plants from the Willowmore district brought into cultivation by Anthony Mitchell have very short, faintly spotted, dark green leaves. The apices are often truncate. They flower during October and November.

'Gamka' Jan Vlok recently collected a very attractive, small form of var. *brachyphylla* from near 'Die Hel' in the Great Swartberg mountain range. It is 110 mm tall, with leaves that have broad truncate margins 10 mm in diameter, giving the appearance of a double keel. It flowers during September and October.

VOUCHERS

Herre 5650 (BOL); *Kramer 68* (PRE); *Lavranos 21019* (NBG); *Moffett 414* (STE); *Van Jaarsveld 7598* (NBG).

12b. **Gasteria brachyphylla** var. **bayeri** *E.J. v. Jaarsveld*: 20 (1992). Type: Cape Province, 10 km south of Calitzdorp, *Bayer 1751* (NBG, holo.!).

HISTORICAL BACKGROUND

Apparently first collected in March 1932 by H.H. Bolus, the variety *bayeri* is named in honour of Bruce Bayer, former curator of the Karoo National Botanical Garden, who has made a substantial contribution to our knowledge of Karoo flora.

DESCRIPTION

Leaves 15-50 mm long, 22-28 mm broad, at first erect, becoming patent and recurved, the apices often incurved, truncate; epidermis smooth or slightly asperulous. *Inflorescence* racemose, 250-280 mm tall. *Perianth* 18 mm long. *Flowering time* October to February.

DISTRIBUTION AND HABITAT

The variety *bayeri* occurs in the Ladismith and Calitzdorp districts of the Little Karoo, growing in succulent karoo, spekboomveld and central mountain renosterveld. It is distributed along the lower slopes of the Swartberg, where it is found at the top of koppies on southern slopes, mainly on shales and rarely on quartzitic sandstones.

Although some rain falls throughout the year, the area is very dry, especially in summer. Plants were seen on top of a koppie among xerophytic vegetation with a preponderance of succulents. The vegetation here consists of low-growing herbs interspersed with larger shrubs.

DIAGNOSTIC FEATURES AND VARIATIONS

As this taxon grades into var. *brachyphylla*, it cannot be accorded any rank other than varietal. It is very variable, especially in leaf size and flowering time. *Gasteria brachyphylla* var. *bayeri* is distinguished from the typical variety mainly by its dwarf stature and obtuse leaves.

VOUCHERS

Bolus s.n. (BOL); *Hayashi 725/84* (NBG); *Laidler 1* (STE); *Van Jaarsveld 7589* (NBG); *Vlok 638* (NBG).

Gasteria brachyphylla var. *bayeri* (all parts life size, except 3 and 4): 1. mature plant; 2. inflorescence; 3. perianth segments (x2); 4. androecium and gynoecium (x3).

Typical *Gasteria bicolor* var. *liliputana* habitat, such as this east of Grahamstown, comprises quartzitic sandstone rock and subtropical thicket.

13. **Gasteria bicolor** *Haw.*: 275 (1825); Haw.: 361 (1827); Bak.: 188 (1880); Bak.: 292 (1896); V. Poelln.: 237 (1938a); Jacobsen: 502 (1954); Jacobsen: 151 (1960). Neo-iconotype: Salm-Dyck: t. 5 (1836-63).

HISTORICAL BACKGROUND

Unfortunately, the well-established name *Gasteria maculata* (Thunb.) Haw. (*Aloe maculata* Haw.) for the above taxon could not be upheld because of the earlier homonym *Aloe maculata* All. (Allioni 1774, see also Dandy 1970). *Gasteria bicolor* Haw., the earliest legitimate name in the synonymy, therefore takes priority.

In 1826 Haworth described *G. bicolor* from live plants introduced by Bowie, but his description was so generalized and lacking in diagnostic detail that it might have referred to a number of *Gasteria* species. In 1827 he subsequently revised and enlarged his description of *G. bicolor*. This last description was considerably more detailed and agrees with an excellent colour plate by Salm-Dyck (1836-63: t. 5). No type material for either description was preserved and this plate was therefore designated as a neotype (Art. 7.9, Greuter *et al.* 1988).

It is not certain who first introduced this species to Europe. Aiton (1789) based his description of *Aloe maculata* var. *obliqua* (*G. bicolor*) on a living plant cultivated by Phillip Miller in 1759. Thunberg collected his *A. maculata* in 1773, but it is doubtful whether he or his companion, Francis Masson, collected live plants or seeds for cultivation.

DESCRIPTION

Plants decumbent to erect, 50-500 mm tall, with a short, leafy stem up to 200 mm tall, proliferating from the base to form small groups. *Roots* succulent, 3 mm in diameter. *Leaves* distichous or spirally arranged, 30-400 mm long, 15-60 mm broad at the base, lorate to linear, erectly spreading, rarely somewhat recurved, slightly falcate and twisted sideways; apices becoming incurved during the dry season; adaxial surface plane to convex, channelled during the dry season; abaxial surface convex when distichous, with an excentric keel when spirally arranged, dark green and with dense white spots arranged in obscure transverse bands, rarely not spotted; epidermis smooth, rarely slightly asperulous to somewhat tuberculate; margin entire, cartilaginous-serrulate, tubercles becoming confluent towards apex, forming an entire semi-translucent white margin; apex obtuse, rarely acuminate, with an excentric mucro. Juvenile leaves patent or erectly spreading, lorate, asperulous; apex obtuse, mucronate. *Inflorescence* 0,1-1,5 m tall, rarely simple or branched from the middle, with up to eight erectly spreading side branches; bracts 10-23 mm long, 3-5 mm broad at the base; floral bracts 3-7 mm long and 1-2 mm broad at the base; pedicel 3-7 mm long. *Perianth* 12-20 mm long, stipitate for 1-2 mm, gasteriform basally for more than half its length, thence abruptly constricted to a tube 3-4 mm wide; gasteriform portion light pink, rarely white, globose to globose-elliptic, inflated to a diameter of 6-9 mm; tube white with green striations; apices of segments erect to erectly spreading, obtuse. *Stamens* 8-17 mm long; anthers oblong, 1,5-3 mm long, not protruding. *Ovary* 4-7 mm long, 1,5-3 mm in diameter; style 6-11 mm long, exceeding the anthers after anthesis; stigma minute, included. *Capsule* 10-25 mm long, 6-10 mm broad. *Seeds* oblong to rectangular, 2-4 mm long. *Flowering time* July to November, with a peak in September and October and sporadically throughout the year.

DISTRIBUTION AND HABITAT

Gasteria bicolor is widely distributed in the eastern Cape, commonly on dry, alluvial coastal plains, in river valleys and on mountainous terrain up to an altitude of 800 m. The species is a typical component of subtropical thicket. In the north-east it also enters the noorsveld and the false karroid brokenveld. The habitat is a dense, thorny, and in some areas impenetrable, sclerophyllous thicket 1-3 m tall, and here *G. bicolor* is prevalent in shady situations together with other succulents. It grows on various rock and soil formations, including conglomerates, shales, mudstones and sandstones, aelionite, quartzitic sandstone and dolerite.

The climate in the habitat is subtropical, but arid. Precipitation, ranging from 300 to 600 mm per annum, occurs in both winter and summer, but the summers tend to be dry in the south-west and the winters dry in the north and east. Temperatures are very hot in summer and moderate in winter; if frost occurs, it is very light.

DIAGNOSTIC FEATURES AND VARIATIONS

Gasteria bicolor is very variable, but is recognized readily by its short, leafy stem which may be up to 200 mm long and the lorate (rarely linear), dark green, slightly falcate leaves which may remain distichous or become spirally arranged (margin serrulate). The leaf apex is almost always obtuse (rarely acute), often twisted sideways with an excentric mucro. It is further characterized by its short perianth, 12-20 mm long, which is basally inflated to a diameter of 9 mm. The inflorescence is a raceme consisting of erectly spreading side branches (excluding var. *liliputana*). *G. bicolor* is very variable in size (30 mm-1,5 m tall). In floral characters it closely resembles *G. baylissiana*, a dwarf acaulescent species with distinctly dense tuberculate leaves. *G. bicolor* is also closely related to *G. brachyphylla* of the Little Karoo. It has similar smooth, dark green leaves and flowers, but *G. brachyphylla* is acaulescent, usually with a less inflated gasteriform perianth base and leaves which often have an acute apex (margin crenulate). Large robust clones of *G. bicolor* var. *bicolor* may also be confused with *G. excelsa*, but the latter is acaulescent with broad, triangular to triangular-lanceolate leaves and has a very large spreading panicle which appears in December.

Two varieties are recognized on the basis of leaf size and inflorescence shape, but they are very variable and grade into each other.

Gasteria bicolor var. *bicolor* (all parts life size, except 3, 4 and 5): 1. mature plant; 2. part of inflorescence; 3. perianth segments and stamens (x2); 4. bract (x2); 5. gynoecium (x4); 6. capsules.

Subtropical thicket on the Ecca Pass near Grahamstown provides shade for *Gasteria bicolor* var. *bicolor*.

13a. **Gasteria bicolor** var. **bicolor**

DESCRIPTION

Leaves 80-220 mm long, distichous or spirally arranged, with or without a distinct excentric keel. *Inflorescence* up to 450 mm tall with two to eight erectly spreading branches.

DISTRIBUTION AND HABITAT (see also under the species)

The variety *bicolor* is widely distributed and frequently encountered throughout the eastern Cape. Near Port Elizabeth plants were found growing in small groups in humus-rich soil on a north-facing slope, in the shade of dense thorny scrub 1-3 m tall. Small trees, shrubs and shrubby climbers are prominent in the vegetation. A distichous form was seen growing at Soutkloof in the shade of dense, thorny scrub that was 1-3 m tall.

DIAGNOSTIC FEATURES AND VARIATIONS

This typical variety is very variable in size and leaf arrangement, but is easily distinguished by its foliate stem up to 200 mm long. The leaves are slightly falcate, usually twisted sideways, with an excentric mucro. Var. *bicolor* differs from var. *liliputana* by its size and paniculate inflorescence. The two varieties grade into each other. A few records exist of plants in the north-eastern part of var. *bicolor's* range that are intermediate between *Gasteria bicolor* and *G. excelsa* and that flower during November. They have short, leafy stems, however, and therefore have been included in *G. bicolor* in this classification. Hybrids have been reported where the habitats of *G. bicolor* and *G. pulchra* overlap.

CULTIVARS

There are a number of local forms, of which the following are worth cultivating:

'*Pearston*' North of the Suurberg range, Grahamstown district, and Klein Winterhoek Mountains, the plants have leaves that remain distichous but the perianth is more inflated than that of plants further south.

'*Mackay Bridge*' This attractive form (previously described as *G. chamaegigas*) was collected by Frank Stayner from Mackay Bridge; it has short leaves arranged in an attractive spiral.

'*Committees Drift*' A very large robust form, previously known as *G. marmorata*, from Committees Drift Bridge, has been brought into cultivation.

'*Bowie*' Bowie's original clonotype of *G. bicolor* has been found in European collections since 1825 (Berger 1908). The plants are slow growing and have attractive, faintly spotted green leaves with an entire margin.

VOUCHERS

Archibald 5971 (PRE); *Galpin 7101* (GRA); *Patterson 95* (BOL); *Thode 706* (NH, PRE); *Van Jaarsveld & Sardien 9070* (NBG).

Gasteria bicolor var. *liliputana*, typical specimen (all parts life size): 1. mature plant; 2. inflorescence; 3. capsules.

Gasteria bicolor var. *liliputana* in a rocky shelter, sharing its habitat with *Crassula lactea*.

13b. **Gasteria bicolor** var. **liliputana** *(V. Poelln.) E.J. v. Jaarsveld*: 21 (1992). *G. liliputana* V. Poelln.: 159 (1938b). Type: Cape Province, Peddie road, 32 km from Grahamstown, *Dyer & Britten 508* (PRE, holo.!).

HISTORICAL BACKGROUND

The variety *liliputana* was first collected by Dr R.A. Dyer and Miss L. Britten in August 1926 (Pole Evans 1929).

DESCRIPTION

Leaves distichous or spirally arranged, 15-100 mm long, 8-14 mm broad at base; epidermis smooth, rarely slightly tuberculate or slightly asperulous; apex obtuse or acute. *Inflorescence* a simple raceme, 160-400 mm long, occasionally with a pair of side branches. *Perianth* 12-15 mm long, diameter of gasteriform portion variable. *Capsule* 12-14 mm long. *Flowering time* July to November, with a peak in September, and occasionally throughout the year.

DISTRIBUTION AND HABITAT

The variety *liliputana* is confined to the Grahamstown and Alicedale districts of the eastern Cape, to the shade or semi-shade of subtropical thicket, on rocky, Palaeozoic quartzitic sandstone ridges and low mountains. The soil is well drained and rich in humus. Plants are usually confined to cooler southern slopes but may also occur on northern slopes in shade. The climate is hot in summer with cool nights, and winters are mild and frost-free where the plants occur.

Some plants have been observed in subtropical thicket on the southern slope of a rocky outcrop. The area is subject to frequent fog, the rocks are lichen-covered, and three epiphytic orchid species were also seen there. The variety *liliputana* occurs in shallow rocky pockets in acid, humus-rich soil. It is fairly abundant, growing together with other succulents, trees and shrubs.

Other plants of the variety *liliputana* have been observed on the northern slope of a gorge near Swartwaterspoort. They occur in dense groups in humus-rich soil and exhibit signs of extensive predation, thought to be by the rock dassie (*Procavia capensis*). Usually only the crown section of the plant is eaten, and new plants proliferate from the broken leaf pieces.

DIAGNOSTIC FEATURES AND VARIATIONS

The variety *liliputana* displays a great deal of variation, especially in size, colour and leaf phyllotaxy. It is distinguished from the typical variety mainly by its dwarf size and simple inflorescence. In other respects it is very similar to the typical var. *bicolor*. Plants of one population of var. *liliputana* east of Grahamstown varied from 25 to 100 mm in height. The plants usually proliferate from the base to form small groups, but a few solitary individuals have also been observed. A solitary clone had very dark, faintly spotted leaves (not due to phenotypic variation). Plants with distichous as well as spirally arranged leaves were seen. This population seems to indicate a high mutation tempo.

CULTIVARS

The following plants have been brought into cultivation:

'*Swartwater*' At Swartwaterspoort most plants have short distichous leaves, but in a few individuals the leaves are spirally arranged. The leaf texture varies from smooth to slightly asperulous. Doreen Court was responsible for bringing this form into cultivation.

'*Pluto's Vale*' A very small form (plants 15-30 mm tall) was introduced by Colonel C.L. Scott from Pluto's Vale.

VOUCHERS

Court 448 (NBG); *Dyer 508* (GRA); *Hardy 2192* (PRE); *Smith 3842* (NBG); *Van Jaarsveld 9095* (NBG).

Gasteria bicolor var. *liliputana*, atypical specimen (all parts life size, except 3, 4 and 5): 1. mature plant; 2. part of inflorescence; 3. perianth (x2); 4. perianth segments, androecium and stipe (x2); 5. gynoecium (x3).

Gasteria baylissiana grows in the shade of subtropical thicket near the top of the Suurberg in the eastern Cape. It is far less conspicuous than the grey-foliaged cycad *Encephalartos lehmannii*.

14. **Gasteria baylissiana** *Rauh*: 187 (1977); Bayer: 29 (1979); Court: 177 (1981); E.J. v. Jaarsveld: 4 (1983). Type: Cape, south of Somerset East, *Bayliss 2796* (HEID, holo.; PRE, iso.!).

HISTORICAL BACKGROUND

Gasteria baylissiana was discovered in 1960 by John Truter, who sent plants to Brand van Breda of the Worcester Veld Reserve from where the plants were distributed. The species was named in honour of Colonel Roy Bayliss, a well-known succulent authority who collected plants for distribution to other botanical gardens in South Africa and elsewhere throughout the world.

DESCRIPTION

Plants acaulescent, decumbent to erect, 15-40 mm high, prolific from the base to form small dense groups up to 80 mm in diameter, rarely solitary. *Roots* succulent, up to 4 mm in diameter. *Leaves* distichous, 25-55 mm long and 20-23 mm broad at the base, lorate, erectly spreading, often becoming patent or recurved (the apices incurved during the dry season); adaxial surface plane or convex; abaxial surface convex; epidermis with dense white cartilaginous tubercles; tubercles very dense, domed to spherical and confluent, forming a dense reticulation; margin crenulate, becoming continuous towards apex; apex obtuse, truncate or retuse, mucronate. Juvenile leaves lorate, tuberculate and slightly asperulous. *Inflorescence* racemose, 80-350 mm tall, erectly spreading; occasionally with a pair of side branches, bearing 15 to 25 flowers; scape 3-4 mm in diameter at the base, reddish pink; bracts 5 mm long, 1 mm broad at the base; pedicel 5 mm long. *Perianth* 14-16 mm long, stipitate for 1 mm, gasteriform basally for two-thirds of its length, thence abruptly constricted to a tube 3-4 mm wide; gasteriform portion red-pink, 6-7,5 mm in diameter; tube white with green striations, 5 mm long; segments free for 5 mm at apex, greenish white with dark green median striations, apices erectly spreading, obtuse; inner perianth segments canaliculate for the length of the gasteriform portion. *Stamens* included; anthers oblong, 1,5 mm long. *Ovary* 4 mm long, 1,75 mm in diameter; style 5 mm long; stigma capitate, included. *Capsule* 14-17(-20) mm long. *Seeds* oblong, 4 mm long, 3 mm wide. *Flowering time* September to October.

DISTRIBUTION AND HABITAT

Gasteria baylissiana is difficult to find in the habitat and is known only from a single farm that lies at the top of the Suurberg range, though it may well have a wider distribution. The plants are confined to an upper eastern slope at an altitude of 750-800 m, growing on quartzitic sandstone. The rocky krantz where the plants are found is densely overgrown with subtropical thicket, but the habitat also contains elements of other veld types. The plants occur singly or in small dense groups, wedged into the vertical rocky crevices, as well as on a steep hill in dense shade below the rock face. They share the habitat with other succulents, for the soil is rich in humus and well drained.

DIAGNOSTIC FEATURES AND VARIATIONS

Gasteria baylissiana is a beautiful, slow-growing, dwarf species, characterized by its densely tuberculate, lorate, erectly spreading leaves and its reddish pink, much inflated gasteriform perianth. It is prolific from the base and forms small dense groups, but new plants will also proliferate from leaf fragments lying on the ground. The leaf apex is obtuse or retuse. The plants are variable in tubercle density as well as leaf size. Plants in light shade or sunny situations have short compact leaves, but those in dense shade develop longer, linear, lorate leaves. The tubercles on the leaves are cartilaginous, white and translucent and very dense in plants occurring in sunny situations. Apart from the phenotypic plasticity, plants also exhibit some genetic variation. In 1977 Rauh reported a more vigorous form with elongated leaves of up to 100 mm long, producing up to three inflorescences a season. In cultivation plants soon adapt to the availability of light.

In floral characteristics *G. baylissiana* closely resembles *G. bicolor* var. *liliputana* which occurs on the same mountain chain, but the leaves of the latter are smooth, usually linear, lanceolate, and spirally arranged. In the Suurberg, however, Mr Tom Hool found an interesting form of *G. bicolor* var. *liliputana* with short, smooth distichous leaves, somewhat reminiscent of *G. baylissiana*. The latter is also related to *G. glomerata*, another dwarf species that forms clusters. Its leaves are, however, glaucous and unspotted, and its leaf epidermis minutely tuberculate. It also has larger flowers.

Superficially *G. baylissiana* also resembles *G. nitida* var. *armstrongii* and small forms of *G. carinata* var. *verrucosa*, but the latter two species are distinguished by their much larger, differently shaped perianths.

VOUCHERS

Bayliss 1330 (NBG, PRE); *Truter s.n.* (NBG); *Van Jaarsveld, Truter & Sardien 9058* (NBG).

Gasteria baylissiana (all parts life size, except 2, 3 and 5): 1. mature cluster of plants; 2. longitudinal sections through perianth showing androecium and gynoecium (x3); 3. gynoecium (x5); 4. capsule; 5. seed (x2).

The rugged environs of the Kouga Dam in the south-eastern Cape provide a habitat for *Gasteria glomerata*. The plants grow in dense clusters on sheer, south-facing cliffs.

15. **Gasteria glomerata** *E.J. v. Jaarsveld*: 100 (1991c). Type: Cape Province, west of Patensie, *Van Jaarsveld & Sardien 11054* (NBG, holo!).

HISTORICAL BACKGROUND

Gasteria glomerata was discovered by the author at two sites in the Kouga Dam area, while exploring the sheer, south-facing cliff faces.

DESCRIPTION

Plants acaulescent, decumbent to erect, 15-40 mm tall, 20-50(-80) mm in diameter, proliferating from the base to form dense clusters up to 200 mm in diameter. *Roots* up to 2 mm in diameter. *Leaves* distichous, 15-25(-50) mm long and 15-20(-25) mm broad at the base, firm, lorate to widely ovate; inner leaves erectly spreading, outer leaves patent or recurved, biconvex in cross section to almost terete, becoming dorsiventrally flattened during the dry season; upper surface often retuse in upper third; both surfaces glaucous, immaculate; epidermis tuberculate-asperulous, usually covered with dust; the tubercles translucent and smooth on top, domed, often confluent and forming a rugulose pattern; outer cell walls between tubercles irregularly domed to subconical, with centric translucent papillae resulting in the asperulous texture; margin entire, minutely crenulate-tuberculate in upper quarter; apex truncate or obtuse, mucronate. Juvenile leaves distichous, lorate, ascending at first, becoming patent or recurved; epidermis asperulous, only slightly and obscurely tuberculate. *Inflorescence* an erectly spreading raceme, 120-200 mm long, bearing 8 to 20 flowers; scape 3 mm in diameter at base; bracts 4-5 mm long, 2-2,5 mm broad at the base; floral bracts the same; pedicel 5-7 mm long. *Perianth* (20-)22-25(-27) mm long, trichromatic, stipitate for 5 mm, gasteriform basally for slightly more than half its length; the gasteriform portion reddish pink, variable, 6-9(-10) mm in diameter (globose-elliptic to globose), thence constricted into a tube 4 mm wide; apices erect to erectly spreading, obtuse, white with centric green striations; inner segments canaliculate for the length of the gasteriform portion. *Stamens* 18-20 mm long, three lengthening in advance; anthers 2 mm long, oblong, included. *Ovary* 4-5 mm long and 2 mm in diameter; style 10-11 mm long, exceeding the anthers after anthesis; stigma minute, curved upwards, becoming capitate during anthesis, included. *Capsule* 13-15 mm long, 6 mm in diameter. *Seeds* 2-3 mm long, 1,5-2 mm wide. *Flowering time* September.

DISTRIBUTION AND HABITAT

Gasteria glomerata is endemic to the Kouga Dam area in the south-eastern Cape, and its discovery came as a surprise. (*G. ellaphieae* from the same area is described on page 50.) The terrain is rugged and inhospitable, and plants occur on sheer, vertical, shady, south-facing rocky ledges (altitude 500-700 m), in slightly acid, quartzitic sandstone soils that are poor in minerals. The plants grow in small or larger dense clusters, exposed or in the shade of other cremnophilous trees or shrubs. *G. glomerata* occurs in the centre of *Gasteria* species diversity and in the vicinity of *G. ellaphieae*, an unrelated species. The two species do not flower simultaneously and no hybrids have been observed.

The rugged habitat of *G. glomerata* is rich in succulent and geophytic plant species. The vegetation in the region consists of subtropical thicket, with fynbos on the upper slopes. The climate is hot in summer and mild in winter. Rainfall, between 300 and 400 mm annually, occurs in summer and winter, but the winters tend to be drier.

DIAGNOSTIC FEATURES AND VARIATIONS

Gasteria glomerata is an attractive dwarf species immediately recognized by its immaculate glaucous leaves, the epidermis of which is tuberculate-asperulous. The distichous leaves are compact, lorate to widely ovate, and recurved. The plants proliferate from the base to form dense, tight clusters and with age become globose in shape. The flowers are bright red-pink, often with a markedly swollen gasteriform portion, in size not exceeded by any other *Gasteria* species. A feature unique among *Gasteria* species is that the capsule often remains pendulous.

G. glomerata is related to two other cremnophilous species, *G. rawlinsonii* and *G. baylissiana*, but differs from them in several ways. Florally *G. glomerata* appears to resemble most closely *G. rawlinsonii*, but vegetatively the two cannot be confused; *G. rawlinsonii* is distinguished by its long, pendent stems bearing asperulous (non-tuberculate) leaves. In vegetative features *G. glomerata* most closely resembles *G. baylissiana*, another dwarf species.

The latter has dark green, densely tuberculate (not asperulous), erectly spreading to patent leaves and a smaller perianth of 14-16 mm. A special feature of these three species is their flowers: the gasteriform portion of the perianth is often markedly globose and varies considerably. The latter is often triangular in cross section, a unique synapomorphy only rarely shared by *G. bicolor*. The distribution of *G. glomerata*, *G. rawlinsonii* and *G. baylissiana* is clearly allopatric and confined to rocky quartzitic sandstone soils. The flowering times of *G. baylissiana* and *G. rawlinsonii* are also in spring.

Another species which has distichous tuberculate leaves with which *G. glomerata* may possibly be confused is *G. nitida* var. *armstrongii*. This taxon has dark green, tuberculate leaves and a perianth of which the gasteriform portion is narrowly elliptic.

The specific epithet *glomerata* (glomerate = compactly clustered) refers to the plant's habit.

VOUCHERS

Van Jaarsveld & Sardien 11054 (NBG).

Gasteria glomerata (all parts life size, except 3): 1. mature plant in flower; 2. infructescence; 3. seed (x5).

Gasteria rawlinsonii on a sheer cliff face in the Baviaanskloof, south-eastern Cape.

16. **Gasteria rawlinsonii** *Oberm.*: t. 1701 (1976); Barkhuizen: 189, t. 181 (1978); Bayer: 30 (1979); Court: 177, t. 176 (1981). Type: Cape Province, Willowmore district, *S.I. Rawlinson s.n. in* PRE *34421* (PRE, holo.!).

HISTORICAL BACKGROUND

Gasteria rawlinsonii was first collected and grown by Marlene Kotze in her garden on the farm Gruisbult. It was seen here by S.I. Rawlinson who, in 1972, brought it to the attention of the botanical world.

DESCRIPTION

Plants caulescent, pendulous, prolific from the base, with a leafy stem up to 1 m long, rarely branched; internodes 10 to 20 mm apart. *Roots* succulent, up to 3 mm in diameter. *Leaves* distichous or spirally arranged, 30-80 mm long and 10-25 mm broad at the base, linear, lorate, slightly falcate; apices recurved; adaxial surface plane to convex; abaxial surface convex, without a keel; both surfaces green, not spotted or with faint white spots; epidermis asperulous; margin sparingly denticulate, or sometimes unarmed (prickles green or white, often turning black with age); prickles 1-2 mm long, 2-4 mm apart; apex obtuse, mucronate. *Inflorescence* racemose, 100-306(-500) mm long, bearing up to 20 reddish pink flowers; bracts 5 mm long, 2 mm broad at the base; pedicel 10 mm long. *Perianth* variable, 16-25 mm long, stipitate for 1-3 mm, gasteriform basally for more than half its length (globose-elliptic or globose), thence constricted to a tube 4-6 mm in diameter; the gasteriform portion pink, 6-9 mm in diameter; tube pink or white, occasionally with green striations. *Stamens* 17 mm long; filaments slightly flattened, becoming contracted after anthesis; anthers 2 mm long, oblong, not protruding. *Ovary* oblong-ovoid, 5 mm long and 2-3 mm in diameter; style 11-12 mm long, exceeding the anthers after anthesis; stigma minute, included. *Capsule* 18 mm long, oblong-ovoid. *Seeds* 3-4 mm wide. *Flowering time* August to October, with a peak in September, and occasionally at other times of the year.

DISTRIBUTION AND HABITAT

Gasteria rawlinsonii is endemic to the mountainous Baviaanskloof region in the south-eastern Cape. At present it is known only from a range about 30 km long, but may well have a wider distribution. The plants occur in steep, shady, narrow ravines, growing in abundance but well out of reach in crevices and on ledges of vertical quartzitic sandstone faces. The soil is rich in humus and is slightly acid. At the foot of the ravine the rocky and sandy terrain is permanently moist, supporting numerous trees and shrubs which form subtropical thicket. The vegetation on the sheer rock faces consists mainly of succulents.

DIAGNOSTIC FEATURES AND VARIATIONS

Gasteria rawlinsonii is most distinctive, both in floral and vegetative characters. Its long, leafy, pendent stems and its oblong-ovoid ovary, which may bear three distal appendages, immediately distinguish it from any other species. In floral characteristics it seems to be allied to *G. glomerata*, *G. bicolor* and *G. baylissiana*. The differences between *G. rawlinsonii* and *G. glomerata* are discussed under the latter species. *G. bicolor* has a short, decumbent to erect, leafy stem up to 200 mm long, and *G. baylissiana* is a dwarf, acaulescent species.

Unique features of *G. rawlinsonii* are the almost immaculate, asperulous leaves which are armed with distinct white- to green prickles that harden and turn black with age. It is also the only *Gasteria* species that does not proliferate from broken leaf pieces. The plants are variable, especially in leaf arrangement and the size of the teeth on the leaf margin. Different populations have either distichous or spirally arranged leaves, and these do not occur together in the same ravine, although they are sometimes separated by only a few kilometres.

CULTIVARS

The following cultivars have been brought into cultivation:

'*Geelhoutboskloof*' Forms from Geelhoutboskloof have leaves which are spirally arranged and have large prickles that turn black with age. This cultivar bears a lax inflorescence.

'*Gert Smitskloof*' Plants introduced from Gert Smitskloof in the Baviaanskloof have attractive compact distichous leaves with few prickles. They bear short, compact inflorescences.

'*Staircase*' Plants introduced by Gerrie Rossouw from an unknown origin in the Baviaanskloof have attractive, spirally arranged distichous leaves, almost like a staircase.

VOUCHERS

De Bruyn s.n. (PRE); *Rawlinson s.n.* (PRE, NBG); *Rossouw 548* (NBG); *Van Jaarsveld & Kotze 7139* (NBG); *Viviers 45/83* (NBG).

Gasteria rawlinsonii (all parts life size): 1. mature plant in flower; 2. distichous form; 3. infructescence.

Appendix 1

INSUFFICIENTLY KNOWN SPECIES

Aloe acinacifolia var. *minor* Salm-Dyck: 13 (1817). Described from a living plant received from a Mr van Marum; no specimen traced.

Aloe formosa Roem. & Schult. 7: 662 (1829). Apparently no specimen preserved; described from a living plant. The description is lacking in diagnostic detail and the name could be applied to a number of species.

Aloe guttata Salm-Dyck: 332 (1834). *Gasteria nigricans* var. *guttata* (Salm-Dyck) Bak.: 185 (1880). Apparently no specimen preserved; the name could be applicable to *G. bicolor* as well as *G. brachyphylla*.

Aloe obliqua Jacq. 4: 418 (1804). Apparently no specimen preserved. The description and illustration might be applicable to *Gasteria bicolor*.

Aloe pseudoangulata Salm-Dyck: 16 (1817). Apparently no specimen preserved; described from a living plant. The description could fit more than one species, but seems to resemble *Gasteria carinata*.

Aloe pseudo-nigricans Salm-Dyck: 17 (1817). Apparently no specimen preserved; described from a living plant. The description suggests that the plant resembles *Gasteria brachyphylla*, but the leaves are slightly tuberculate.

Aloe vittata Roem. & Schult. 7: 662 (1829). Apparently no specimen preserved; described from a living plant. The description could be applied to *Gasteria bicolor* as well as *G. brachyphylla*.

Gasteria acinacifolia var. *spathulata* Berger: 158 (1908). Described from a living plant in the Kew collection. The type in the Kew Herbarium (K, holo.!) appears to be of hybrid origin.

Gasteria angulata var. *truncata* (Willd.) Berger: 134 (1908). *Aloe angulata* var. *truncata* Willd.: 16 (1811). Apparently no specimen preserved. The description could be applied to *G. disticha*, *G. brachyphylla* or *G. carinata*.

Gasteria brevifolia Haw.: 97 (1819). Apparently no specimen preserved; described from a living plant. The description could be applied to *G. brachyphylla*, *G. disticha* or *G. carinata*.

Gasteria brevifolia var. *laete-virens* Haw.: 351 (1827). Apparently no specimen preserved; described from a living plant. The description could be applied to a number of species.

Gasteria brevifolia var. *perviridis* Haw.: 351 (1827). Apparently no specimen preserved; described from a living plant. The description could be applied to a number of species.

Gasteria crassifolia Haw.: 351 (1827). *Aloe nigricans* var. *crassifolia* Salm-Dyck: 17 (1817). Apparently no specimen preserved; described from a living plant. The description could be applied to a number of species.

Gasteria laevis (Salm-Dyck) Haw.: 355 (1827). *Aloe laevis* Salm-Dyck: 16 (1817). *A. angustifolia* var. *laevis* Salm-Dyck: 58 (1821). Apparently no specimen preserved; described from a living plant. The name could be applied to *G. brachyphylla* or *G. carinata* var. *carinata*.

Gasteria elongata Bak.: 290 (1896). Described from a living plant in the Kew collection in 1885. The description could be applied to a number of species.

Gasteria fasciata (Salm-Dyck) Haw.: 350 (1827). Apparently no specimen preserved; described from a living plant. *Aloe nigricans* var. *fasciata* Salm-Dyck: 64 (1821). *G. nigricans* var. *fasciata* (Salm-Dyck) Haw.: 47 (1821). The description could be applied to either *G. brachyphylla* or *G. bicolor*.

Gasteria fasciata var. *laxa* Haw.: 350 (1827). Apparently no specimen preserved; described from a living plant. *Aloe vittata* var. *laxa* (Haw.) Roem. & Schult. 7: 663 (1829). The description could be applied to *G. brachyphylla* or *G. bicolor*.

Gasteria fasciata var. *polyspila* (Bak.) Berger: 140 (1908). *G. nigricans* var. *polyspila* Bak.: 185 (1880). Apparently no specimen preserved; described from a living plant. The description could be applied to more than one species.

Gasteria formosa Haw.: 350 (1827). Apparently no specimen preserved; described from a living plant. The description is lacking in diagnostic detail and could fit a number of taxa.

Gasteria gracilis Bak.: 192 (1880). Apparently no specimen preserved; described from a living plant. The description does not mention floral parts and does not fit any known species.

Gasteria linita Haw.: 361 (1827). Described from a living plant in the Kew collection. An early illustration of this taxon (one of the Kew portraits) is lacking in diagnostic detail and the name could not be applied with certainty to any specific taxon.

Gasteria prolifera Lemaire: 58 (1869). Apparently no specimen preserved. The description could not be applied with certainty to any known *Gasteria* species.

Gasteria subnigricans Haw.: 351 (1827). *Aloe subnigricans* (Haw.) Roem. & Schult. 6: 17 (1829). *G. nigricans* var. *subnigricans* (Haw.) Bak.: 186 (1880). Apparently no specimen preserved. The description suggests that the plant closely resembles *G. brachyphylla*, but the leaves are slightly tuberculate.

Gasteria subnigricans var. *canaliculata* Salm-Dyck 29: t. 10 (1836-1863). Apparently no specimen preserved; the description could be applied to a number of *Gasteria* species and may represent a hybrid.

Gasteria subnigricans var. *glabrior* Haw.: 352 (1827). Apparently no specimen preserved. The description could not be applied with certainty to any *Gasteria* taxon.

Gasteria transvaalensis Bak.: 44 (1889). Apparently no specimen preserved; described from a living plant. The description could be applied to *G. disticha* or *G. brachyphylla*.

Gasteria trigona (Salm-Dyck) Haw.: 358 (1827). *Aloe trigona* Roem. & Schult. 7: 678 (1829); *A. trigona* var. *elongata* Salm-Dyck: 45 (1821); *A. elongata* Salm-Dyck 29: t. 15 (1836-1863); Kunth 4: 539 (1843); *A. acinacifolia* var. *angustifolia* Salm-Dyck: 13 (1817); *A. acinacifolia* var. *laete-virens* Salm-Dyck: 13 (1817). Salm-Dyck based his *A. trigona* var. *elongata* on the above two varieties of his *A. acinacifolia*. Both were described from living plants and the names could be applied to a number of taxa.

Gasteria trigona var. *kewensis* Berger: 153 (1908). Described from a living plant. The description could be applied to a number of taxa.

Appendix 2

PUTATIVE HYBRIDS

Gasteria apricoides Bak.: 197 (1880). Apparently no specimen preserved; described from a living plant.

Gasteria bayfieldii (Salm-Dyck) Bak.: 197 (1880). Apparently no specimen preserved; described from a living plant. *Aloe bayfieldii* Salm-Dyck 29: t. 14 (1836-1863).

Gasteria cheilophylla Bak.: 189 (1880). Apparently no specimen preserved; described from a living plant.

Gasteria holtzei (Radl) Berger: 131 (1908). *Aloe holtzei* Radl: 24 (1896). Type: unknown.

Gasteria kewensis Berger: 131 (1908). Type: cultivated at Kew, apparently no specimen preserved.

Gasteria lauchei (Radl) Berger: 131 (1908). *Aloe lauchei* Radl: 27 (1896). Type: unknown.

Gasteria margaretifera Berger: 131 (1908). Type: cultivated at La Mortola, no specimen preserved.

Gasteria metallica Berger: 153 (1908). Type: cultivated at La Mortola, no specimen preserved.

Gasteria peacockii Bak.: 195 (1880). Described from a living plant raised by Pfersdorff in Paris about 1875.

Gasteria penthamensis Bak.: 193 (1880). Type: garden hybrid between *G. carinata* and *Aloe variegata*, raised at Pentham, Kent, in 1819 by a Mr Ricketts.

Gasteria rufescens Berger: 138 (1908). Type: cultivated at Riviera Garden, apparently no specimen preserved.

Gasteria squarrosa Bak.: 197 (1880). Apparently no specimen preserved; described from a living plant in the Kew collection

Appendix 3

CULTURAL USES AND VERNACULAR NAMES

FOOD

Gasteria flowers are edible and were first eaten by the Khoi (Smith 1966). When eaten raw, they have a flavour reminiscent of a sweet fresh bean though with a burning after-taste. Smith (1966) reported that the young buds of *Gasteria brachyphylla* were boiled as a rice by the Khoi, hence the vernacular name 'Hottentot rice', and in the Robertson and Oudtshoorn districts the flowers of *Gasteria disticha* are used as a vegetable by farmers (Pepler 1980), who call it 'oukossie'. Stripped off the scape, the flowers are prepared in a *bredie* (stew).

According to Mrs S. Marais of the farm Die Vlakte in Robertson, an oukossie stew is traditionally prepared during the September-October wine festival which coincides with the *Gasteria disticha* flowering period. It is prepared in the following manner:

In 2 dessertspoons of oil, fry 2 coarsely chopped large onions until soft. Add 1 kg of fatty sheep rib cut into cubes and fry until brown. Next add 1,5 cups of boiling water and cook the meat and onion mixture until it is nearly dry. Boil 750 g of freshly gathered and washed gasteria flowers for 3 to 5 minutes, then drain. Season with salt and pepper, and sprinkle generously with lemon juice. Cut up 4 medium potatoes and add to the meat with the gasteria flowers. Cook for a further 40 minutes.
(Pers. comm. Mrs S. Marais, Die Vlakte, Robertson.)

MEDICINAL USE

The Xhosa consider the leaves of *Gasteria croucheri* to be a remedy for paralysis. The leaves, pulped in cold water, are used to wash a patient's body (Watt & Breyer-Brandwijk 1962).

'MAGICAL' USES

In rural South Africa, plants are still commonly used for 'magical' purposes. A specific feature of a plant or animal is often believed to be acquired by the user when ingested in a particular way. For instance, the heart of a lion would render strength, or the eye of a vulture keen eyesight. The mottled leaves of *Gasteria* species provide a camouflage effect so that it blends into its background, making it difficult to find. F.A. van Jaarsveld jnr (pers. comm.) relates that for this reason *G. croucheri* is highly prized for its magical properties by Zulu warriors in Natal, who believe that, when eaten, it will make the user partially invisible. It is often consumed before faction fights or in an attempt to hide from an enemy.

Gasteria is known to the Nguni as 'ntelez' (that is, a plant with magical properties). The Xhosa believe that when placed on top of their homes, the plants will ward off lightning. Occasionally plants can be found actually growing on the rooftops of Transkei dwellings.

VERNACULAR NAMES

Vernacular names as related by Smith (1966):

'Bontalwee' (*bont* = mottled, *aalwyn* = aloe) is a name used to describe various *Gasteria* species.

'Bosaalwee' (*bos* = bush) has been used for *G. acinacifolia* and refers to the habitat where this species grows.

'Beestong' (= ox tongue) is used for *G. disticha* and it refers to the strap-shaped leaves.

'Bontbesie' (*besie* = little animal) is the name given to *G. disticha* in the southern parts of the Great Karoo.

GASTERIA SYNONYMS

Aloe acinacifolia Jacq. var. *venusta* (Haw.) Salm-Dyck: 29 (1836-1849) = *Gasteria acinacifolia.*

Aloe acinacifolia Jacq.: 49, t. 31 (1811-1816); Ker: t. 2369 (1823); Salm-Dyck: 12 (1817); Roem. & Schult.: 679 (1829); Salm-Dyck: t. 11 (1836-1863); Kunth: t. 4 (1843) = *Gasteria acinacifolia.*

Aloe angulata Willd.: 276 (1811) = *Gasteria carinata* var. *carinata.*

Aloe angustifolia Salm-Dyck: 57 (1821) = *Gasteria disticha.*

Aloe boureana Roem. & Schult.: 662 (1829) = *Gasteria bicolor* var. *bicolor.*

Aloe bowieana Salm-Dyck: t. 3 (1836-1863) = *Gasteria bicolor* var. *bicolor.*

Aloe brachyphylla Salm-Dyck: t. 8 (1840, see Stafleu & Cowan 1979), cited 1836-1863 = *Gasteria brachyphylla* var. *brachyphylla.*

Aloe candicans (Haw.) Roem. & Schult.: 681 (1829) = *Gasteria acinacifolia.*

Aloe carinata DC.: t. 63 (1799) = *Gasteria carinata* var. *verrucosa.*

Aloe carinata Mill.: n. 1 (1768); Haw.: 13 (1804); Ker: t. 1331 (1811); Salm-Dyck: 20 (1836-1863); Kunth: 541 (1843) = *Gasteria carinata* var. *carinata.*

Aloe carinata var. *laevior* Salm-Dyck: 29 (1836-1849) = *Gasteria carinata* var. *carinata.*

Aloe carinata var. *subglabra* Haw.: 14 (1804) = *Gasteria carinata* var. *carinata.*

Aloe conspurcata Salm-Dyck: 18 (1817) = *Gasteria disticha.*

Aloe crassifolia Roem. & Schult.: 664 (1829) = *Gasteria disticha.*

Aloe croucheri Hook. f.: t. 5812 (1869) = *Gasteria croucheri.*

Aloe decipiens Roem. & Schult.: 674 (1829) = *Gasteria nitida* var. *nitida.*

Aloe dictyodes Roem. & Schult.: 663 (1829); Salm-Dyck: t. 4 (1836-1863) = *Gasteria bicolor* var. *bicolor.*

Aloe disticha L.: 321 (1753) *pro parte*; L.: 459 (1762) *pro parte* = *Gasteria carinata* var. *carinata.*

Aloe disticha L.: 321 (1753); Roem. & Schult.: 666 (1829) = *Gasteria disticha.*

Aloe disticha L.: 459 (1762); *pro parte* (var. *beta*) = *Gasteria disticha.*

Aloe disticha Thunb.: 7 (1785) non L. (1753) = *Gasteria carinata* var. *verrucosa.*

Aloe ensifolia (Haw.) Roem. & Schult.: 687 (1829) = *Gasteria acinacifolia.*

Aloe excavata Willd.: 276 (1811) = *Gasteria carinata* var. *carinata.*

Aloe glabra Salm-Dyck: 13 (1817) = *Gasteria carinata* var. *carinata.*

Aloe intermedia Haw.: 12 (1804) = *Gasteria carinata* var. *verrucosa.*

Aloe intermedia var. *asperrima* Salm-Dyck: 69 (1821) = *Gasteria carinata* var. *verrucosa.*

Aloe laetepuncta (Haw.) Roem. & Schult.: 676 (1829) = *Gasteria carinata* var. *carinata.*

Aloe lingua Ker: t. 1322 (1810) = *Gasteria carinata* var. *verrucosa.*

Aloe lingua Ker: t. 979 (1807) = *Gasteria bicolor* var. *bicolor.*

Aloe lingua var. *angulata* Haw.: 13 (1804) = *Gasteria carinata* var. *carinata.*

Aloe lingua var. *angustifolia* Ait.: 469 (1789) = *Gasteria disticha.*

Aloe lingua var. *crassifolia* Ait.: 469 (1789) = *Gasteria disticha.*

Aloe lingua var. *latifolia* Haw.: 13 (1804) = *Gasteria disticha.*

Aloe lingua var. *longifolia* Haw.: 13 (1804) = *Gasteria disticha.*

Aloe lingua var. *multifaria* Haw.: 13 (1804) = *Gasteria carinata* var. *carinata.*

Aloe linguiformis Mill.: no. 13 (1768); DC.: 68 (1799) = *Gasteria disticha.*

Aloe maculata Ker: t. 765 (1805a), non Allioni = *Gasteria pulchra.*

Aloe maculata Thunb. var. *pulchra* Ait.: 469 (1789) = *Gasteria pulchra.*

Aloe maculata Thunb.: 8 (1785); Salm-Dyck: t. 1 (1836-1863) = *Gasteria bicolor* var. *bicolor.*

Aloe maculata var. *obliqua* Ait.: 469 (1789) = *Gasteria bicolor* var. *bicolor.*

Aloe nigricans Haw.: 13 (1804) = *Gasteria disticha.*

Aloe nigricans var. *denticulata* Salm-Dyck: 17 (1817) = *Gasteria disticha.*

Aloe nigricans var. *marmorata* Salm-Dyck: 64 (1821) = *Gasteria brachyphylla* var. *brachyphylla.*

Aloe nitens (Haw.) Roem. & Schult.: 680 (1829) = *Gasteria acinacifolia.*

Aloe nitida Salm-Dyck: 13 (1817); Salm-Dyck: 47 (1821); Ker: t. 2304 (1822); Salm-Dyck: t. 17 (1836-1863); Roem. & Schult.: 677 (1829); Kunth: 539 (1843) = *Gasteria nitida* var. *nitida.*

Aloe nitida var. *grandipunctata* Salm-Dyck: 47 (1821) = *Gasteria nitida* var. *nitida.*

Aloe nitida var. *major* Salm-Dyck: 13 (1817) = *Gasteria nitida* var. *nitida.*

Aloe nitida var. *minor* Salm-Dyck: 13 (1817) = *Gasteria nitida* var. *nitida.*

Aloe nitida var. *obtusa* Salm-Dyck: 13 (1817) = *Gasteria nitida* var. *nitida.*

Aloe nitida var. *parvipunctata* Salm-Dyck: 47 (1821) = *Gasteria nitida* var. *nitida.*

Aloe obliqua DC.: t. 91 (1802, cited as 1799), non Haw. = *Gasteria pulchra.*

Aloe obliqua Haw.: 14 (1804) = *Gasteria bicolor* var. *bicolor.*

Aloe obliqua var. *fallax* (Haw.) Roem. & Schult.: 661 (1829) = *Gasteria bicolor* var. *bicolor.*

Aloe obscura var. *truncata* Salm-Dyck: 15 (1817) = *Gasteria carinata* var. *carinata.*

Aloe obscura Willd.: 275 (1811). *Gasteria latifolia* (Haw.) Haw.: 87 (1812) = *Gasteria disticha.*

Aloe obtusa (Salm-Dyck) Roem. & Schult.: 679 (1829) = *Gasteria nitida* var. *nitida.*

Aloe obtusifolia Salm-Dyck: 62 (1821); Salm-Dyck: t. 37 (1836-1863) = *Gasteria disticha.*

Aloe planifolia Bak.: t. 162 (1870) = *Gasteria bicolor* var. *bicolor.*

Aloe pluripuncta (Haw.) Roem. & Schult.: 681 (1829) = *Gasteria acinacifolia.*

Aloe pulchra (Ait.) Jacq.: t. 419 (March 1805, cited as 1804) = *Gasteria pulchra.*

Aloe racemosa Lam.: 91 (1783) = *Gasteria carinata* var. *verrucosa.*

Aloe scaberrima Salm-Dyck: 322 (1834) = *Gasteria carinata* var. *verrucosa.*

Aloe subcarinata Salm-Dyck: 14 (1817) = *Gasteria carinata* var. *carinata.*

Aloe subverrucosa Salm-Dyck: 18 (1817) = *Gasteria carinata* var. *verrucosa.*

Aloe subverrucosa var. *grandipunctata* Salm-Dyck: 67 (1821) = *Gasteria carinata* var. *verrucosa.*

Aloe subverrucosa var. *parvipunctata* Salm-Dyck: 67 (1821) = *Gasteria carinata* var. *verrucosa.*

Aloe sulcata Salm-Dyck: 54 (1821) = *Gasteria carinata* var. *carinata.*

Aloe trigona var. *obtusa* Salm-Dyck: 46 (1821) = *Gasteria nitida* var. *nitida.*

Aloe tristicha Medik.: 74 (1786) = *Gasteria carinata* var. *carinata.*

Aloe venusta (Haw.) Roem. & Schult.: 681 (1829) = *Gasteria acinacifolia.*

Aloe verrucosa Mill.: 20 (1768); Willd.: 189 (1799); Ker: t. 837 (1805b); Salm-Dyck: 19 (1817); Salm-Dyck: t. 25 (1836-1863); Kunth: 543 (1843) = *Gasteria carinata* var. *verrucosa.*

Aloe verrucosa var. *latifolia* Salm-Dyck: 72 (1821) = *Gasteria carinata* var. *verrucosa.*

Aloe verrucosa var. *striata* Salm-Dyck: 19 (1817) = *Gasteria carinata* var. *verrucosa.*

Aloe verrucula Medik.: 63 (1784) = *Gasteria carinata* var. *verrucosa.*

Aloe, section *Curviflorae* Haw.: 12 (1804); Salm-Dyck: 11 (1817); Salm-Dyck: 29 (1836-1863) = *Gasteria.*

Basionym: *Gasteria liliputana* V. Poelln.: 159 (1938b); V. Poelln.: 3, t. 26 (1938c); Phillips in Pole Evans: t. 350 (1929) non *Gasteria pulchra* (Ait.) Haw.; Jacobsen: 635 (1954) = *Gasteria bicolor* var. *liliputana.*

Gasteria acinacifolia (Jacq.) Haw. var. *ensifolia* (Haw.) Bak.: 196 (1880) = *Gasteria acinacifolia.*

Gasteria acinacifolia (Jacq.) Haw. var. *pluripuncta* (Haw.) Bak.: 300 (1896) = *Gasteria acinacifolia.*

Gasteria acinacifolia (Jacq.) Haw. var. *venusta* (Haw.) Bak.: 300 (1896) = *Gasteria acinacifolia.*

Gasteria acinacifolia var. *nitens* (Haw.) Bak.: 196 (1880); Berger: 157 (1908); V. Poelln.: 235 (1938a); Jacobsen: 625 (1954) = *Gasteria acinacifolia.*

Gasteria angulata (Willd.) Haw.: 354 (1827) non Haw. (1812) = *Gasteria carinata* var. *carinata*.
Gasteria angulata Haw.: 88 (1812) = *Gasteria carinata* var. *carinata*.
Gasteria angustianum V. Poelln.: 1 (1937a) = *Gasteria brachyphylla* var. *brachyphylla*.
Gasteria angustifolia (Ait.) Duval: 6 (1809) = *Gasteria disticha*.
Gasteria armstrongii Schönl.: 258 (1912); V. Poelln.: 237 (1938a); Jacobsen: 2626 (1954); Court: 175 (1981) = *Gasteria nitida* var. *armstrongii*.
Gasteria beckeri Schönl.: 140 (1907) = *Gasteria nitida* var. *nitida*.
Gasteria biformis V. Poelln.: 18 (1940) = *Gasteria bicolor* var. *bicolor*.
Gasteria bijliae V. Poelln.: 166 (1937c) = *Gasteria carinata* var. *carinata*.
Gasteria caespitosa V. Poelln.: 165 (1937c) = *Gasteria bicolor* var. *bicolor*.
Gasteria candicans Haw.: 46 (1821) = *Gasteria acinacifolia*.
Gasteria carinata auctt.: Rowley: 1 (1951); non (Mill.) Duval (1809) = *Gasteria batesiana*.
Gasteria carinata var. *falcata* Berger: 150 (1908) = *Gasteria carinata* var. *carinata*.
Gasteria carinata var. *latifolia* Berger: 150 (1908) = *Gasteria carinata* var. *carinata*.
Gasteria carinata var. *parva* (Haw.) Bak.: 296 (1896) = *Gasteria carinata* var. *carinata*.
Gasteria carinata var. *strigata* (Haw.) Bak.: 296 (1896) = *Gasteria carinata* var. *carinata*.
Gasteria chamaegigas V. Poelln.: 159 (1938b) = *Gasteria bicolor* var. *bicolor*.
Gasteria colubrina N.E. Br.: 38 (1877) = *Gasteria bicolor* var. *bicolor*.
Gasteria conspurcata (Salm-Dyck) Haw.: 353 (1827) = *Gasteria disticha*.
Gasteria decipiens Haw.: 357 (1827) = *Gasteria nitida* var. *nitida*.
Gasteria denticulata (Salm-Dyck) Haw.: 50 (1819) = *Gasteria disticha*.
Gasteria disticha var. *angustifolia* (Ait.) Bak.: 187 (1880) = *Gasteria disticha*.
Gasteria disticha var. *conspurcata* (Salm-Dyck) Bak.: 187 (1880) = *Gasteria disticha*.
Gasteria disticha var. *major* Haw.: 352 (1827) = *Gasteria disticha*.
Gasteria disticha var. *minor* Bak.: 187 (1880) = *Gasteria disticha*.
Gasteria disticha var. *minor* Haw.: 352 (1827) = *Gasteria disticha*.
Gasteria disticha var. *natalensis* Bak.: 187 (1880) = *Gasteria croucheri*.
Gasteria ensifolia Haw.: 282 (1825); Haw.: 360 (1827) = *Gasteria acinacifolia*.
Gasteria ernesti-ruschii Dinter & V. Poelln.: 239 (1938d); Jacobsen: 631, t. 537 (1954) = *Gasteria pillansii* var. *ernesti-ruschii*.
Gasteria excavata (Willd.) Haw.: 354 (1827) = *Gasteria carinata* var. *carinata*.
Gasteria fuscopunctata Bak.: 195 (1880) = *Gasteria acinacifolia*.
Gasteria glabra (Salm-Dyck) Haw.: 87 (1812) = *Gasteria carinata* var. *carinata*.
Gasteria herreana V. Poelln.: 154 (1938e) = *Gasteria bicolor* var. *bicolor*.
Gasteria humilis V. Poelln.: 140 (1929) = *Gasteria carinata* var. *carinata*.
Gasteria huttoniae N.E. Br.: 440 (1908) = *Gasteria acinacifolia*.
Gasteria inexpectata V. Poelln.: 45 (1938f) = *Gasteria acinacifolia*.
Gasteria intermedia var. *laevior* Haw.: 356 (1827) = *Gasteria carinata* var. *verrucosa*.
Gasteria intermedia (Haw.) Haw.: 89 (1812) = *Gasteria carinata* var. *verrucosa*.
Gasteria intermedia var. *asperrima* (Salm-Dyck) Haw.: 47 (1821) = *Gasteria carinata* var. *verrucosa*.
Gasteria intermedia var. *longior* Haw.: 356 (1827) = *Gasteria carinata* var. *verrucosa*.
Gasteria joubertii V. Poelln.: 20 (1940) = *Gasteria brachyphylla* var. *brachyphylla*.
Gasteria kirsteana V. Poelln.: 15 (1940) = *Gasteria bicolor* var. *bicolor*.
Gasteria laetepuncta Haw.: 357 (1827) = *Gasteria carinata* var. *carinata*.
Gasteria longiana V. Poelln.: 156 (1938e) = *Gasteria bicolor* var. *bicolor*.
Gasteria longibracteata V. Poelln.: 59 (1938f); V. Poelln.: 33 (1939a) = *Gasteria bicolor* var. *bicolor*.
Gasteria longifolia (Haw.) Duval: 6 (1809) = *Gasteria disticha*.
Gasteria lutzii V. Poelln.: 242 (1933) = *Gasteria acinacifolia*.
Gasteria maculata (Thunb.) Haw.: 349 (1827) = *Gasteria bicolor* var. *bicolor*.
Gasteria maculata var. *dregeana* Berger: 122 (1908); V. Poelln.: 2 (1938g) = *Gasteria bicolor* var. *bicolor*.
Gasteria maculata var. *fallax* Haw.: 349 (1827) = *Gasteria bicolor* var. *bicolor*.
Gasteria marmorata Bak.: 194 (1880) = *Gasteria bicolor* var. *bicolor*.
Gasteria mollis Haw.: 46 (1821). *Aloe mollis* Roem. & Schult.: 655 (1829); Salm-Dyck: t. 38 (1836-1863) = *Gasteria disticha*.
Gasteria multiplex V. Poelln.: 16 (1940) = *Gasteria bicolor* var. *bicolor*.
Gasteria neliana V. Poelln.: 139 (1930) = *Gasteria pillansii* var. *pillansii*.
Gasteria nigricans (Haw.) Duval: 6 (1809) = *Gasteria disticha*.
Gasteria nigricans var. *crassifolia* (Ait.) Haw.: 47 (1821) = *Gasteria disticha*.
Gasteria nigricans var. *marmorata* Haw.: 47 (1821) = *Gasteria brachyphylla* var. *brachyphylla*.
Gasteria nigricans var. *platyphylla* Bak.: 186 (1880) = *Gasteria brachyphylla* var. *brachyphylla*.
Gasteria nitens Haw.: 48 (1819) = *Gasteria acinacifolia*.
Gasteria nitida var. *grandipunctata* (Salm-Dyck) Berger: 152 (1908); V. Poelln.: 243 (1938a); Jacobsen: 640 (1954) = *Gasteria nitida* var. *nitida*.
Gasteria nitida var. *parvipunctata* (Salm-Dyck) Berger 33: 152 (1908) = *Gasteria nitida* var. *nitida*.
Gasteria obliqua (Haw.) Duval: 6 (1809). Haw.: 86 (1812) = *Gasteria bicolor* var. *bicolor*.
Gasteria obtusa (Salm-Dyck) Haw.: 359 (1827) = *Gasteria nitida* var. *nitida*.
Gasteria obtusifolia (Salm-Dyck) Haw.: 351 (1827) = *Gasteria disticha*.
Gasteria pallescens Bak.: 190 (1880) = *Gasteria carinata* var. *carinata*.
Gasteria parva Haw.: 356 (1827) = *Gasteria carinata* var. *carinata*.
Gasteria parvifolia Bak.: 193 (1880) = *Gasteria carinata* var. *carinata*.
Gasteria patentissima V. Poelln.: 19 (1940) = *Gasteria carinata* var. *carinata*.
Gasteria picta Haw.: 349 (1827) = *Gasteria bicolor* var. *bicolor*.
Gasteria planifolia (Bak.) Bak.: 188 (1880) = *Gasteria bicolor* var. *bicolor*.
Gasteria pluripuncta Haw.: 360 (1827) = *Gasteria acinacifolia*.
Gasteria poellnitziana Jacobsen 2: 641, t. 544 (1954), sine descr. Lat. = *Gasteria pulchra*.
Gasteria porphyrophylla Bak.: 190 (1880) = *Gasteria carinata* var. *carinata*.
Gasteria radulosa Bak.: 43 (1889) = *Gasteria carinata* var. *verrucosa*.
Gasteria repens Haw.: 48 (1821) = *Gasteria carinata* var. *verrucosa*.
Gasteria retata Haw.: 350 (1827); Bak.: 189 (1880) = *Gasteria bicolor* var. *bicolor*.
Gasteria salmdyckiana V. Poelln.: 57 (1938f); V. Poelln. 35 (1939a) = *Gasteria bicolor* var. *bicolor*.
Gasteria schweickerdtiana V. Poelln.: 245 (1938a) = *Gasteria carinata* var. *carinata*.
Gasteria spiralis Bak.: 189 (1880) = *Gasteria bicolor* var. *bicolor*.
Gasteria spiralis var. *tortulata* Bak.: 189 (1880); Bak.: 295 (1896) = *Gasteria bicolor* var. *bicolor*.
Gasteria stayneri V. Poelln.: 19 (1938h) = *Gasteria nitida* var. *nitida*.
Gasteria strigata Haw.: 357 (1827) = *Gasteria carinata* var. *carinata*.
Gasteria subcarinata (Salm-Dyck) Haw.: 49 (1819); Haw.: 358 (1827) = *Gasteria carinata* var. *carinata*.
Gasteria subverrucosa (Salm-Dyck) Haw. var. *marginata* auctt.: Hall: 63 (1951); non Bak. (1880) = *Gasteria batesiana*.
Gasteria subverrucosa (Salm-Dyck) Haw.: 355 (1827) = *Gasteria carinata* var. *verrucosa*.
Gasteria subverrucosa var. *grandipunctata* (Salm-Dyck) Haw.: 355 (1827) = *Gasteria carinata* var. *verrucosa*.
Gasteria subverrucosa var. *marginata* Bak.: 184 (1880) = *Gasteria carinata* var. *verrucosa*.
Gasteria subverrucosa var. *parvipunctata* (Salm-Dyck) Haw.: 355 (1827) = *Gasteria carinata* var. *verrucosa*.

Gasteria sulcata (Salm-Dyck) Haw.: 354 (1827) = *Gasteria carinata* var. *carinata*.
Gasteria thunbergii N.E. Br.: 140 (1923) = *Gasteria carinata* var. *verrucosa*.
Gasteria transvaalensis auctt. Oberm.: t. 1501 (1967); Barkhuizen: 190, t. 182 (1978); Court: 177 (1981); non Bak. in *J. Bot.*, Lond. 27: 44 (1889) = *Gasteria batesiana*.
Gasteria triebneriana V. Poelln.: 158 (1938b) = *Gasteria brachyphylla* var. *brachyphylla*.
Gasteria variolosa Bak. 4: t. 347 (1871) = *Gasteria bicolor* var. *bicolor*.
Gasteria venusta Haw.: 360 (1827) = *Gasteria acinacifolia*.
Gasteria verrucosa var. *scaberrima* (Salm-Dyck) Bak.: 288 (1896) = *Gasteria carinata* var. *verrucosa*.
Gasteria verrucosa var. *asperrima* (Salm-Dyck) V. Poelln.: 249 (1938a) = *Gasteria carinata* var. *verrucosa*.
Gasteria verrucosa var. *intermedia* (Haw.) Bak.: 184 (1880) = *Gasteria carinata* var. *verrucosa*.
Gasteria verrucosa var. *latifolia* (Salm-Dyck) Haw.: 47 (1821) = *Gasteria carinata* var. *verrucosa*.
Gasteria verrucosa var. *striata* (Salm-Dyck) V. Poelln.: 249 (1938a) = *Gasteria carinata* var. *verrucosa*.
Gasteria vlaaktensis V. Poelln.: 21 (1940) = *Gasteria brachyphylla* var. *brachyphylla*.
Gasteria zeyheri (Salm-Dyck) Bak.: 190 (1880). *Aloe zeyheri* Salm-Dyck: t. 3 (1836-1863) = *Gasteria bicolor* var. *bicolor*.
Haworthia nigricans Haw.: 301 (1824) = *Gasteria nitida* var. *nitida*.

Glossary

abaxial: describing the lower surface of a leaf, away from the axis or stem
acaulescent: stemless
acropetally: describing of the production of flowers in succession, the oldest at the base and the youngest at the tip
acute: sharply pointed
acuminate: tapering to a point
adaxial: describing the upper surface of a leaf, facing the axis or stem
adpressed: lying flat for the whole length of the organ
allopatric: describing taxa (species) that are geographically separated
aloine: chemical substance commonly found in the family Asphodelaceae
amplexicaul: having a base that encircles the stem
anther: end part of the stamen which contains the pollen
anthesis: the time at which a flower comes into full bloom
apex (-ices/-ical): pointed end or tip
apomorphy: derived or advanced condition of character
arborescent: having tree-like characteristics
arcuate: curved
asperulous: rough, with short hard tubercles due to outgrowths of the outer cell walls
autapomorphy: a unique character state
axil: the angle between leaf and branch
axillary: arising from an axil
bract: a small, leaf-like structure subtending the pedicel or inflorescence branch
canaliculate: channelled
capitate: head-like
cartilaginous: hard and tough, like cartilage in texture
caulescent: with a stem
cladistics: a method of classification by which individuals are grouped according to evolutionary relationships
cladogram: branching diagram of a putative evolutionary tree
clavate: club-shaped
cline: character gradient, the continuous variation of a species character
clonotype: clones (or exact replicas) taken from the original type plant on which the species was based
cremnophilous: cliff-dwelling
crenulate: having a margin finely notched with rounded projections
decumbent: lying flat with the end curving up
dehiscing: splitting into parts
dendrogram: branching diagram depicting relationships
denticulate: very finely toothed
dimorphic: describing the occurrence of two distinct forms of any plant
discrete: separate
distal: at the end furthest from attachment
distichous: describing leaves oppositely arranged in two vertical rows
dorsifixed: describing an anther attached on the dorsal side
emarginate: notched at the abrupt or rounded apex
endemic: confined to a specific area
ensiform: shaped like a sword
entire: describing a margin that is continuous and not broken into teeth or lobes
epidermis: outer cell layer of a leaf
epipeltate: like a shield, with an attachment near the top
equilaterally: applied to flowers that open more or less at the same lateral plane
excentric: off-centre, one-sided
exsert: to thrust, to protrude; to project beyond the corolla of a flower
falcate: shaped like a sickle
filament: the stalk of a stamen
filiform: thread-like
fusiform: (of succulent roots) tapering towards each end
gasteriform: term derived from the generic name *Gasteria*, from the Greek word *gaster* = belly or stomach; referring to the basally inflated perianth
genotype: a group of organisms with the same genetic constitution
geophyte: a perennial plant that propagates by means of buds below the soil surface
glabrous: without hair or similar growth; smooth
glaucous: grey-green
globose: almost spherical
heterogeneous: not uniform
holotype: the type specimen, designated to carry the name of a new species
homologous: having the same evolutionary origin, but not necessarily the same structure and/or function
homoplasy: non-homologous similarity due to convergence or parallelism
hypogynous: inserted below the ovary
iconotype: an illustration accompanying the original description when no type specimen was designated
infructescence: inflorescence in a fruiting stage
introgression: hybridization of two taxa resulting in the spread of genes of one taxon into the gene pool of another
introrse: (of an anther) turning towards the axis; pertaining to an anther dehiscing inwards
isotype: duplicate of the holotype
karroid: karoo-like
karyotypic: applied to the character of a cell nucleus as defined by the shape and number of chromosomes
keel: a narrow ridge on the undersurface of a leaf, like the keel of a boat
koppie: a South African term referring to a rocky hillside

krantz: a South African term referring to a sheer cliff face
lanceolate: spear-shaped, tapering to a pointed apex
lax: having loosely arranged parts
lectotype: the specimen or illustration selected as the type specimen by the subsequent taxonomist when the original author used more than one specimen in his description
linear: parallel-edged, much longer than broad
locular: divided into compartments by partitions
loculicidal: describing a capsule which splits at the midrib of the carpels
lorate: strap-shaped
maculate: spotted or blemished
marginiform: like a margin
mesic: of, relating to, or growing in conditions of medium water supply
mesophyll: the photosynthetic tissue between the upper and lower epidermis of a leaf
monophyletic: derived from a common ancestor
morphology: the study of external shape and development
mucro: a short sharp point
nectary: the organ below the ovary that produces nectar
neo-endemic: a recent endemic
neotype: the specimen or illustration selected when no specimens or illustrations were designated by the original author
obclavate: club-shaped but reversed, the thickened end at the base
obligate: able to exist under only one set of environmental conditions
oblong: longer than broad, with sides more or less parallel
obtuse: rounded or blunt at the end
ontogeny: the origin and development of the individual
ornithophilous: pollinated by birds
outbreeding: cross-breeding between populations or species
ovate: egg-shaped in outline (two-dimensional)
ovoid: egg-shaped, the broader end at the base (three-dimensional)
panicle: a branched raceme with stalked flowers
papilla: a protuberance, or outgrowth on the epidermis
parallelism: the independent attainment of a character state by related taxa (not due to common inheritance)
paratype: a dried plant specimen representing a series of the original plant collection, but not chosen by the author as the type specimen
parenchyma: isodiametric cells of the leaf mesophyll
patent: (of leaves) spread out widely
pedicel: the stalk of a single flower
pendulous: hanging, suspended
perianth: floral leaves surrounding the stamens and carpels
phenetic tree: a tree-like diagram depicting relationships without regard to the evolutionary history
phenotypic: pertaining to the physical constitution of an organism as determined by the interaction of its genetic constitution and the environment
phloem pole: the end of a vascular bundle
phyllotaxy: leaf arrangement on the stem
phylogeny: the genealogical history of a group of organisms
phytophagous: feeding on plants
piliferous: ending in a hair-like point
placentation: the way in which the ovules are attached to the ovary
plesiomorphic: pertaining to the ancestral condition or character state
proliferate: reproduce itself by vegetative multiplication
prolific: producing offshoots abundantly
propagule: a small side shoot that becomes detached from the parent plant and ultimately develops into a mature plant
protandrous: said of flowers in which the anthers ripen before the stigma
putative: considered to exist or have existed; commonly regarded as being
raceme: an inflorescence with a lengthened axis bearing stalked flowers, the oldest flower at the base
recurved: curved or bent backwards
restiads: reed-like plants belonging to the family Restionaceae and typical representatives of the fynbos flora
retuse: notched at the abrupt or obtuse apex
reversal: a return to the ancestral condition
rosulate: bearing leaves in a rosette
rugulose: wrinkled
scape: a leafless flower stalk
scarious: dry and membranous
sclerenchyma: a tissue composed of thick-walled cells
sclerophyll: a firm, leathery leaf consisting of a large amount of sclerenchyma, characteristic of many woody plants adapted to hot, dry areas
secund: pertaining to flowers turned in the same direction
self-incompatible: self-sterile, ie the ripe pollen is not effective on its own stigma, so there is no seed set
serrate: notched on the edge, like a saw
serrulate: minutely notched
sister group: close relatives derived from the same immediate ancestor
solitary: single
stigma: at the end of the style, the part that is receptive to pollen
stipitate: stalked
stolon: a basal sucker or runner that ultimately forms a new plant
striations: narrow parallel bands of colour
style: the upper part of the pistil, between the ovary and stigma
subconical: almost cone-shaped
subulate: awl-shaped
sympatric: pertaining to species that occur together
sympodial: pertaining to growth (the main axis stops growing and the lateral bud develops)
synapomorphy: the common possession of an inherited homologous character
taxonomy: the classification of (plant) life
terete: cylindrical or tapering
topocline: series of gradual variations within a species throughout a geographical range
trichromatic: involving three colours
trilobate: with three apical lobes
truncate: ending abruptly
tuberculate: covered with tubercles
type (specimen): the herbarium specimen on which the description of a species is based
undulating: wavy
vascular bundle: a specialized bundle of small conducting vessels in a plant, consisting of the xylem and phloem
velamen: water-absorbing tissue on the outside of roots
ventricose: swollen
versatile: describing an anther that turns freely on the apex of the filament
vicariant: pertaining to two closely related, but geographically isolated species.
xeromorphic: having characteristics that serve as protection against loss of water
xerophyte: a plant adapted to growing in dry conditions

Acocks, J.P.H. 1953. Veld types of South Africa. *Memoirs of the Botanical Survey of South Africa* no. 28. Government Printer, Pretoria.
Aiton, W. 1789. *Hortus Kewensis*, edn 1. Nicol, London.
Allioni, C. 1774. Auctarium ad synopsim methodicam stirpium horti reg. Taurinensis. *Miscellanea Taurinensia* 5: 53-96.
Baker, J.G. 1870. *Aloe planifolia*, t. 162. *Saunders Refugium Botanicum*.
Baker, J.G. 1871. *Gasteria variolosa*, t. 347. *Saunders Refugium Botanicum*.
Baker, J.G. 1880. A synopsis of Aloineae and Yuccoideae. *Journal of the Linnean Society*, Botany 18: 148-241.
Baker, J.G. 1889. New petaloid monocotyledons from Cape Colony. *Journal of Botany* 27: 44.
Baker, J.G. 1896. *Gasteria*. In W.T. Thiselton-Dyer, *Flora capensis* 6: 286-307. L. Reeve, London.
Barkhuizen, B. 1978. *Succulents of South Africa*. Perskor, Johannesburg.
Bayer, M.B. 1979. *Gasteria*. *Excelsa* 9: 29-33.
Bayer, M.B. 1982. *The new* Haworthia *Handbook*. National Botanical Gardens of South Africa, Cape and Transvaal Printers, Cape Town.
Berger, A. 1908. *Liliaceae-Asphodeloideae-Aloineae*. In A. Engler, *Das Pflanzenreich* 33: 122-160. Leipzig.
Bewley, J.D. and Black, M. 1986. *Seeds, physiology of development and germination*. Plenum Press, New York and London.
Boerhaave, H. 1720. *Index alter plantarum quae in horto academico Lugduno - Batavo aluntur*. Van der As, Leiden.
Bolus, H. and Kensit, L. 1909. Contributions to the African flora. *Transactions of the Royal Society of South Africa* 1: 163.
Bond, P. and Goldblatt, P. 1984. Plants of the Cape flora. *Journal of South African Botany*, supplementary volume 13. Trustees of the National Botanic Gardens of South Africa, Kirstenbosch.
Bowie, J. 1830. Miscellaneous intelligence (Botany). *The South African Quarterly Journal* 1: 90-91.
Brewbaker, J.L. and Gorrez, D.D. 1967. Genetics of self-incompatibility in the monocot genera, *Ananas* (pineapple) and *Gasteria*. *American Journal of Botany* 54: 611-616.
Brown, N.E. 1877. *Gasteria colubrina*. *Gardener's Chronicle* 8: 38.
Brown, N.E. 1908. LV. - *Diagnoses Africanae*. *Kew Bulletin* 26: 440.
Brown, N.E. 1923. The genera *Aloe* and *Mesembryanthemum* as represented in Thunberg's Herbarium. *Bothalia* 1: 140.
Commelin, C. 1701. *Horti medici amstelodamensis rariorum plantarum descriptio et icones*. Fredericus Haringh, Leiden.
Court, G.D. 1981. *Succulent flora of southern Africa*. A.A. Balkema, Cape Town.
Cowling, R.M. 1982. Patterns of plant endemism in the south-eastern Cape. *The Naturalist* 27: 17-36.
Cronquist, A. 1981. *An integrated system of classification of flowering plants*. Columbia University Press, New York.
Cutler, D.F. and Brandham, P.E. 1977. Experimental evidence for the genetic control of leaf surface characters in hybrid Aloineae (Liliaceae). *Kew Bulletin* 32: 23-32.
Dahlgren, R.M.T. and Clifford, H.T. 1982. *The Monocotyledons, a comparative study*. Academic Press, London.
Dahlgren, R.M.T., Clifford, H.T. and Yeo, P.F. 1985. *The families of the Monocotyledons*. Springer-Verlag, Berlin, Heidelberg, New York, Tokyo.
Dandy, E. 1970. Annotated list of the new names published in Allioni's *Auctarium ad synopsis meth. stirpium Hort. reg. Taurinensis*. *Taxon* 19: 617-618.
De Candolle, A.P. 1799-1823. *Plantarum succulentarum historia ou histoire naturelle des Plantes Grasses*. Dugour & Durand, Paris.
Drège, J.F. 1843. *Zwei pflanzengeographische Dokumente*. Leipzig.
Duval, H.A. 1809. *Plantae succulentae in Horto Alenconio*. Gabon, Paris.
Dyer, R.A. 1976. *Genera of southern African flowering plants*, Vol. 2, Gymnosperms and Monocotyledons. Government Printer, Pretoria.
Gibbs Russell, G.E. and Robinson, E.R. 1981. Phytogeography and speciation in the vegetation of the eastern Cape. *Bothalia* 13: 467-472.
Goldblatt, P. 1978. An analysis of the flora of South Africa; its characteristic relationships and origins. *Annals of the Missouri Botanical Gardens* 65: 369-436.
Greuter, W. (editor), Burdet, H.M., Chaloner, W.G., Demoulin, V., Grolle, R., Hawksworth, D.L., Nicholson, D.H., Silva, P.C., Stafleu, F.A., Voss, E.G. 1988. *International code of botanical nomenclature*. Adopted by the Fourteenth International Botanical Congress, Berlin, July-August 1987. *Regnum vegetabile*, a series of publications for the use of plant taxonomists published under the auspices of the International Association for Plant Taxonomy. Bohn, Scheltema & Holkema, Utrecht.
Gunn, M.D. and Codd, L.E. 1981. *Botanical exploration of southern Africa*. A.A. Balkema, Cape Town.
Hall, A.V., De Winter, M., De Winter, B. and Van Oosterhout, S.A.M. 1980. *Threatened plants of southern Africa*. South African National Scientific Programmes Report, no. 45. Pretoria.
Hall, H. 1951. Hall Marks. *National Cactus and Succulent Journal* 6: 63.
Haworth, A.H. 1804. A new arrangement of the genus *Aloe*. *Transactions of the Linnean Society of London* 7: 1-28.
Haworth, A.H. 1812. *Synopsis plantarum succulentarum*. Taylor, London.
Haworth, A.H. 1819. *Supplementum plantarum succulentarum*. Harding, London.
Haworth, A.H. 1821. *Revisiones plantarum succulentarum*. Wood, London.
Haworth, A.H. 1824. *Haworthia nigricans*. *Philosophical Magazine* 62: 301.
Haworth, A.H. 1825. Description of new succulent plants. *Philosophical Magazine* 46: 282-283.
Haworth, A.H. 1826. Description of new succulent plants. *Philosophical Magazine* 57: 275.
Haworth, A.H. 1827. Description of new succulent plants. *Philosophical Magazine* 2: 344-361.
Herre, H. 1971. *The genera of the Mesembryanthemaceae*. Tafelberg, Cape Town.
Hooker, J.D. 1869. *Aloe croucheri*. *Curtis's Botanical Magazine* 95: t. 5812.
Hutchinson, J. 1934. *The families of flowering plants*. 2: Monocotyledons. Macmillan & Company Ltd, London.
Jacobsen, H. 1954. *Handbuch der sukkulenten Pflanzen* 2: 624-648. Fischer, Jena.
Jacobsen, H. 1960. *A handbook of succulent plants*, 2: 500-520. Blandford Press Ltd, London.
Jacobsen, H. 1974. *Lexicon of succulent plants*, 237-244. Blandford Press Ltd, London.
Jacobsen, H. and Rowley, G.D. 1955. Some name changes in succulent plants - Part 1. *National Cactus & Succulent Journal* 10,2: 32.
Jacquin, N.J. Von. 1804. *Plantarum rariorum horti caesarei schoenbrunnensis descriptiones et icones*. Vienna.
Jacquin, J.F. Von. 1811-1844. *Eclogae plantarum rariorum aut minus cognitarum quas ad vivim descripsit et iconibus coloratis illustravit*. The Author, Vienna.
Kamstra, M.W. 1969. *Aloe* seeds. *Aloe* 6,2: 8-10.
Kamstra, M.W. 1971. *Aloe* seeds. *Excelsa* 1: 19-26.
Karsten, M.C. 1951. *The Old Company's Garden at the Cape and its Superintendents*. Maskew Miller, Cape Town.
Ker, J.B. 1805a. *Aloe maculata* Ker. *Curtis's Botanical Magazine* 20: t. 765.

Ker, J.B. 1805b. *Aloe verrucosa* Ker. *Curtis's Botanical Magazine* 22: t. 837.

Ker, J.B. 1807. *Aloe lingua* Ker. *Curtis's Botanical Magazine* 25: t. 979.

Ker, J.B. 1810. *Aloe lingua* Ker. *Curtis's Botanical Magazine* 33: t. 1322.

Ker, J.B. 1811. *Aloe carinata* Ker. *Curtis's Botanical Magazine* 33: t. 1331.

Ker, J.B. 1822. *Aloe nitida* Ker. *Curtis's Botanical Magazine* 49: t. 2304.

Ker, J.B. 1823. *Aloe acinacifolia* Ker. *Curtis's Botanical Magazine* 50: t. 2369.

Kunth, C.S. 1843. *Enumeratio plantarum*. 534-549. Colla, Stuttgart.

Lamarck, J.B.A.P.M. de. 1783-1817. *Encyclopedie methodique* (Dictionnaire encyclopedique de botanique). Panckoacke Paris.

Lemaire, C. 1869. *Les Plantes Grasses autres que les Cactees* 3: 54-59. Maison rustique, Paris.

Linder, H.P. 1983. The historical phytogeography of the Disinae (Orchidaceae). *Bothalia* 14, 3: 565-570 (1983).

Linnaeus, C. 1753. *Species plantarum*. Exhibentes plantas rite cognitas genera relatas ... etc. Salvius, Stockholm.

Linnaeus, C. 1762. *Species plantarum*, 2nd edn. Exhibentes plantas rite cognitas genera relatas ... etc. Salvius, Stockholm.

Maire, R. 1958. *Gasteria*. In *Flore de l'Afrique du nord* 5: 71-75. Monocotyledonae: Liliales: Liliaceae. Lechevalier, Paris.

Marshak, A. 1934. Chromosomes and compatibility in the Aloineae. *American Journal of Botany* 21: 592-598.

Medikus, F.K. 1784. *Botanische Beobachtungen des Jahres* 1783. Hofbuchhandlung, Mannheim.

Medikus, F.K. 1786. *Theodora speciosa*. Hofbuchhandlung, Mannheim.

Miller, P. 1755-1760. *Figures of the most beautiful, useful and uncommon plants described in the Gardener's Dictionary, with the characters of their flowers and seed vessels etc.* Rivington, London.

Miller, P. 1768. *The Gardener's Dictionary, containing the methods of cultivating and improving the kitchen, fruit, flower garden and nursery*, 8th edn. Rivington, London.

Obermeyer, A.A. 1967. *Gasteria transvaalensis*. *The Flowering Plants of Africa* 38: t. 1501.

Obermeyer, A.A. 1976. *Gasteria rawlinsonii*. *The Flowering Plants of Africa* 43: t. 1701.

Pepler, D. 1980. Oukossie - Klein Karoo-bredie. *Landbouweekblad*, 26 September, 141: 118.

Pole Evans, I.B. 1928. *Gasteria carinata*. *The Flowering Plants of Africa* 8: t. 291.

Pole Evans, I.B. 1929. *Gasteria liliputana*. *The Flowering Plants of Africa* 9: t. 360.

Pole Evans, I.B. 1937. *Gasteria excelsa*. *The Flowering Plants of Africa* 17: t. 665.

Radford, A.E. 1986. *Fundamentals of plant systematics*. Harper & Row, New York.

Radl, F. 1896. Neue *Aloe*-Hybriden des Abts Regium. *Monatschrift für Kakteenkunde* 6: 23-27.

Rauh, W. 1977. *Gasteria baylissiana* Rauh: A new dwarf *Gasteria* from south-east Africa. *Journal of South African Botany* 43 (3): 187-191.

Reynecke, P.R. 1965. *The genus* Astroloba *Uitewaal (Liliaceae)*. M.Sc. thesis, University of Cape Town.

Reynolds, G.W. 1950. *The Aloes of South Africa*. The Trustees of the Aloes of South Africa Book Fund, Johannesburg.

Reynolds, G.W. 1966. *The Aloes of Tropical Africa and Madagascar*. The Trustees of the Aloe Book Fund, Mbabane, Swaziland.

Reynolds, G.W. 1982. *The Aloes of South Africa*, 4th edn. A.A. Balkema, Cape Town.

Riley, H.P. and Madjumdar, S.K. 1979. *The Aloineae, a biosystematic survey*. University Press, Kentucky.

Roemer, J.A. and Schultes, J.H. 1829. *Systema vegetabilium*. 660-679. Cotta, Stuttgart.

Rowley, G.D. 1951. Succulents in the Bates collection. *National Cactus and Succulent Journal* 6,1: 6.

Rowley, G.D. 1967. A numerical survey of the genera of Aloineae. *National Cactus and Succulent Journal* 12: 71-75.

Rowley, G.D. 1976a. Generic concepts in the Aloineae, Part I - The genus - to split or to lump? *National Cactus and Succulent Journal* 31,2: 27-31.

Rowley, G.D. 1976b. Generic concepts in the Aloineae, Part II - Generic names in the Aloineae. *National Cactus and Succulent Journal* 31 3: 54-56.

Rowley, G.D. 1982. Intergeneric hybrids in succulents. *National Cactus and Succulent Journal* 37,3: 76.

Salm [Reifferscheid] Dyck, Prince J.M.F.A.H.I. 1817. *Catalogue raisonne des especes d'Aloes*. Dusseldorf.

Salm [Reifferscheid] Dyck, Prince J.M.F.A.H.I. 1821. *Observations botanicae in horto Dyckensi*. Thiriart, Koln.

Salm [Reifferscheid] Dyck, Prince J.M.F.A.H.I. 1834. *Hortus Dyckensis*. Arne, Dusseldorf.

Salm [Reifferscheid] Dyck, Prince J.M.F.A.H.I. 1836-1863. *Monographia generum Aloes et Mesembryanthemi*. Henry & Cohen, Dusseldorf, Bonn.

Schelpe, E.A.C.L.E. 1958. *Gasteria*, a problem genus of South African succulent plants. *Journal of the Botanical Society of South Africa* 44: 17-20.

Schönland, S. 1907. New and some little known species of South African plants belonging to the genera *Aloe*, *Gasteria*, *Crassula*, *Cotyledon* and *Kalanchoe*. *Records of the Albany Museum* 2: 140.

Schönland, S. 1912. A new species of *Gasteria*. *Records of the Albany Museum* 2: 258.

Smith, C.A. 1966. *Common names of South African plants*. The Government Printer, Pretoria.

Smith, G.F. 1985. *Die taksonomiese status van die monotipiese genusse* Chortolirion *Berger en* Poellnitzia *Uitewaal in die tribus Aloineae (Fam. Asphodelaceae)*. Submitted in partial fulfilment of the requirement for the degree Baccalaureus Scientiae (Honores), University of Pretoria, Pretoria.

Smith, G.F. 1991. *Contributions to the systematics of selected genera of the Alooideae (Asphodelaceae)*. Submitted in partial fulfilment of the requirements for the degree Doctor of Philosophy in the Faculty of Science (Dept. of Botany), University of Pretoria, Pretoria.

Snijman, D. 1984. *A revision of the genus* Haemanthus. *Journal of South African Botany*, Supplementary Vol. 12.

Stafleu, A. and Cowan, R.S. 1979. *Taxonomic literature*, Vol. 5. Bohn, Scheltema & Holkema, Utrecht.

Stearn, W.T. 1938. An annotated index to Salm-Dyck's *Monographia generum Aloes et Mesembryanthemi*. *Cactus Journal*, December: 34-44.

Stearn, W.T. 1958. Botanical exploration to the time of Linnaeus. *Proceedings of the Linnaean Society, London* 169: 173-193.

Stearn, W.T. 1961. *Botanical gardens and botanical literature in the eighteenth century*. Catalogue of the botanical books in the collection of Rachel McMasters Miller Hunt. 2,1. Pittsburgh. The Hunt Botanical Library, compiled by Alan Stevenson.

Stearn, W.T. 1971. Adrian Haworth. In H. Herre, *The genera of the Mesembryanthemaceae*. Tafelberg, Cape Town.

Thunberg, C.P. 1785. *Dissertationes academicae Upsaliae habitae*. Edman, Uppsala.

Tilli, M.A. and Florentino, C.F. 1723. *Catalogus plantarum horti Pisani*. Florence.

Valentyn, 1726. *Beschryvinge van de Kaap der Goede Hoop*. Thiende Boek, Tweede Hoofstuk 5: 22-23. In: *Beschryvinge van oud en nieuw oost-Indien*. Van Braam, Dordrecht.

Van der Pijl, L. 1982. *Principles of dispersal in higher plants*. Springer-Verlag, Berlin, Heidelberg, New York.

Van der Walt, J.J.A. 1977. *Pelargoniums of southern Africa*. Purnell, Cape Town.

Van der Walt, J.J.A. and Vorster, P.J. 1981. *Pelargoniums of southern Africa*, Vol 2. Juta, Cape Town.

Van der Walt, J.J.A. and Vorster, P.J. 1988. *Pelargoniums of southern Africa*, Vol. 3. *Annals of Kirstenbosch Botanic Gardens*, Vol. 16.

Van Jaarsveld, E.J. 1983. *Gasteria baylissiana* Rauh - 'n aantreklike dwergspesie van die Oos-Kaap. *Aloe* 20, 1: 4-5.

Van Jaarsveld, E.J. 1987a. *Gasteria vlokii*: a new species from the south-east Cape. *Cactus and Succulent Journal (U.S.)* 59: 170-174.

Van Jaarsveld, E.J. 1987b. The succulent riches of South Africa and Namibia. *Aloe* 24, 3 & 4: 46-92.

Van Jaarsveld, E.J. 1989. *A revision of the genus* Gasteria *(Asphodelaceae)*. University of Natal. A thesis submitted for the fulfilment of the requirements for the degree of Master of Science in the Department of Botany, Faculty of Science, University of Natal, Pietermaritzburg.

Van Jaarsveld, E.J. 1990. *Gasteria baylissiana. The flowering plants of Africa* 51,1: t. 2015.

Van Jaarsveld, E.J. 1991a. The conservation status of *Gasteria baylissiana* Rauh, a rare endemic of the Zuurberg (Eastern Cape). *British Cactus and Succulent Journal* 9, 4: 99-101.

Van Jaarsveld, E.J. 1991b. *Gasteria ellaphieae* Van Jaarsveld: a new *Gasteria* species of the south-east Cape. *Cactus and Succulent Journal (U.S.)* 63, 1: 3-7.

Van Jaarsveld, E.J. 1991c. *Gasteria glomerata* Van Jaarsveld sp. nov. *Bradleya* 9: 100-104.

Van Jaarsveld, E.J. 1992. The genus *Gasteria*, a synoptic review. *Aloe* 29: 1-30.

Visser, D.J.L. 1984. *Geological map of the Republics of South Africa, Transkei, Bophuthatswana, Venda and the Kingdoms of Lesotho and Swaziland*. Department of Mineral and Energy affairs. Government Printer, Pretoria.

Von Poellnitz, K. 1929. Zur Kenntnis der Gattungen *Haworthia* Duv. und *Gasteria* Duv. *Feddes Repertorium Specierum Novarum Regni Vegetabilis* 27: 132-143.

Von Poellnitz, K. 1930. *Gasteria neliana. Feddes Repertorium Specierum Novarum Regni Vegetabilis* 27: 139.

Von Poellnitz, K. 1933. *Gasteria lutzii* spec. nov. aus dem Kapland. *Feddes Repertorium Specierum Novarum Regni Vegetabilis* 33: 242-243.

Von Poellnitz, K. 1937a. *Gasteria angustiarum* von Poellnitz spec. nov. *Kakteen und andere Sukkulenten* 10, 2: 1-2.

Von Poellnitz, K. 1937b. Eine seltene *Gasteria*-art. *Kakteen und andere Sukkulenten* 8: 119-120.

Von Poellnitz, K. 1937c. Zwei neue *Gasteria*-arten. *Kakteenkunde* 11: 165-167.

Von Poellnitz, K. 1938a. Zur Kenntnis der Gattung *Gasteria* Duval. *Feddes Repertorium Specierum Novarum Regni Vegetabilis* 63: 233-250.

Von Poellnitz, K. 1938b. Zur Kenntnis der Gattung *Gasteria* Duval. *Feddes Repertorium Specierum Novarum Regni Vegetabilis* 65: 157-160.

Von Poellnitz, K. 1938c. Three new and interesting species of the genus *Gasteria*, Duval. *The Cactus Journal* 7: 3.

Von Poellnitz, K. 1938d. Die erste *Gasteria*-art aus Sudwestafrika. *Gasteria ernesti-ruschii* Dinter et Von Poellnitz spec. nov. *Kakteen und andere Sukkulenten* 63: 239.

Von Poellnitz, K. 1938e. Two new species of the genus *Gasteria*, Duval. *Desert Plant Life* 10: 156.

Von Poellnitz, K. 1938f. Drei neue *Gasteria*-arten, (8). *Feddes Repertorium Specierum Novarum Regni Vegetabilis* 65: 56-60.

Von Poellnitz, K. 1938g. *Gasteria maculata* var. *dregeana* Berger. *The Cactus Journal* 7,2: 58-59.

Von Poellnitz, K. 1938h. *Gasteria stayneri* Von Poellnitz spec. nov. *The Cactus Journal* 7,1: 18-20.

Von Poellnitz, K. 1939a. Three new species of the genus *Gasteria*, Duval. *Desert Plant Life* 11: 33-36.

Von Poellnitz, K. 1939b. *Gasteria nigricans* (Haw.) Duval. *Desert Plant Life* 11: 512-513.

Von Poellnitz, K. 1940. Neue *Gasteria*-arten (3). *Feddes Repertorium Specierum Novarum Regni Vegetabilis* 69: 15-23.

Watt, J.M. and Breyer-Brandwijk, M.G. 1962. *The medicinal and poisonous plants of southern and eastern Africa*. Livingstone, London.

Weimarck, H. 1941. Phytogeographical groups, centres and intervals within the Cape flora. *Lunds Universitets Arsskrift*. NF Avd. 2 Bd., 37,5: 1-143.

Werger, J.A. 1978. *Biogeography and ecology of southern Africa*. Junk, The Hague.

White, F. 1971. The taxonomic and ecological basis of chorology. *Mitteilungen der Botanischen Staatssammlung, München* 10: 91-112.

White, F. 1983. *UNESCO/AETFAT/UNSO vegetation map of Africa*. AETFAT vegetation map Committee, Paris.

Wijnands, D.O. 1983. *The botany of the Commelins*. Balkema, Rotterdam.

Willdenow, C.L. 1799. *Species plantarum*, 4th edn. Nauk, Berlin.

Willdenow, C.L. 1811. Bemerkungen über die Gattung *Aloe* aus *Magazin der Gesellschaft Naturforschender Freunde aus Berlin für die neuesten Entdeckungen in der gesam(m)ten Naturkunde* 5: 163-283.

Index

Page numbers in bold refer to the main description of the species; page numbers in italic indicate an illustration. Synonyms are followed by *.

COLLECTORS' EDITION

Diana L. Barrell
Capt. John D. Baxter, Australia
Jim Gerard Paul Broekhuysen
Charles Craib
Hubert C. de Monmonier
Dr P. K. de Villiers
Dr J. Dhansay
Mr & Mrs J. M. R. Dower
W. T. F. (Will) du Toit
John & Julie Fannin
Fernwood Press

First National Bank
Eugéne & Lalie Fourie
Russel, Bonnie & Gabriella Friedman
Basil Hersov
J. A. Hilhorst
John & Nakkie Hoffman
Rupert & Nadia Horley
Paul Isserow
Kjeld Jacobsen
Edmund Kirschnek
Daryl Koutnik & Rana Tayyar

M. Kristen
O. J. Mackenzie
Harry Chi-king Mak
Carel A. Nolte
SA Library
L. N. Schroder
Pieter Struik
Dr H. J. van Wyk
Ellaphie Ward-Hilhorst
Ilse & Stephan Wentzel
Whitestone Gardens Ltd

STANDARD EDITION

Stanley Peel Abramson
Africana Book Collectors
Alexander L. D. Agenbacht
Dario Alaimo
Albertus Delport-biblioteek
Ian W. Alderton
Stephen Allmand
C. D. Anderson
Erik Anderson
P. Fraga Arguimbau
Etwin Aslander
Alex Austin

Marijke & Ken Ball
Professor D. E. Barclay
Ms Sonia J. Barker-Fricker
Professor B. P. Barkhuizen
S. R. Bateman
Brian Bates, Langold, England
Auriol Batten
Yvonne Becker
Jack & Melanie Beith
John G. Bell
Beatrice Bellegoni, Sarzana, Italy
Keith Benson
Ilse Eckle Benthaus
Ari & Gia Bert
Avv. Guglielmo M. Betto
Professor Gerhard J. Beukes
Lorraine Bielckus
Mike & Trish Bingham
John E. Bishop
James W. Bizzell
Raymond Blomme
H. Bongers
Rob & Kathy Borthwick
Jürgen Bosack

T. F. Bosch
Linda Brack
Steven Brack
Edna & Frank Bradlow
Dieter & Susan Brandt
Ingo Breuer
P. D. Brink
Paul D. Brink
Roy & Joan Brunette
Mrs Christine Bruno
Mrs Margaret Bruty
Pieter Bruwer
Dr H. J. Brynard
Mr W. Bullock
Mnr en Mev. P. M. Burger
Vonkies Burger
Chris & Veronika Buys

Cactus & Succulent Society of Queensland Inc.
Charles Cadman
Geoffrey H. Card
Vincent Cerutti
Elisabeth & Graham Charles
C. F. Clinning
Trevor Coleman
Jean Collier
Maria & Tony Conradie
Bert Coppus
N. R. Court
C. L. Cousins
Alain W. Craig
Dr J. A. Cremona
Edmond Crombez
Pierre & Françoise Cuénod

Walter Dams
Mrs Roslyn E. M. Davey

Stoffel de Jager
K. N. de Kock
Cheryl de Lange
Hubert C. de Monmonier
H. A. C. de Villiers
P. J. de Villiers
Dr Raymond C. Dedeyne
Christian Defferrard
C. J. & M. J. Dekker
Deutsche Kakteen-Gesellschaft
Paul & Monique Dewanckel
Philip E. Downs
Colin F. Drake
Paul E. Drez
Dr J. P. Driver-Jowitt
E. J. du Plessis
Francois du Randt
Johan en Suzette du Toit
W. T. F. (Will) du Toit
Graham Duncan
Ed Dunne & Mary Parisi

Eastern Province Wild Flower Society
L. Edgecombe
John Edmonds
Urs Eggli
David John Ellis
Endangered Wildlife Trust
Mucchiati Ernani
J. M. Esterhuizen
Euclea Books
R. W. H. Evans

Mr I. R. Facey
Kobus Farrell
David Fergobella
J. Barry Ferguson

Meeding & Barendien Ferreira
Liz & Alex Fick
Dr Nathan Finkelstein
Kirby W. Fong
Diane Foster
Malcolm Foster
Beverley Fourie
Eugéne & Lalie Fourie
Robin Frandsen
Mrs J. E. Fraser-Jones
Georg & Hannes Fritz
Barry Frost

Chris Gaigher & Family
J. W. Gamesby
Janet & Antony Gane
Mr Robin C. Gardner
Ian Garland
Geelong Cactus & Succulent Club
François Georges
Dr S. K. Gess
Rupert Gettliffe
Jan Giliomee
R. St. L. Gillman
Michael G. N. Goodwin
Dr G. J. M. Gorter
Keith Grantham
Josephine Gray
D. Green
Mrs M. F. Green
Rodney & Susan Green
Dr Ryno Greenwall
Elizabeth Jane Greenwood
Paul Primich Griessel
Carl Grobbelaar
Charmaine Grobbelaar
N. Grobbelaar
Professor Y. Gutterman
Gerald A. H. Guy

Mr Hall
A. C. J. Hall
R. S. & B. A. Hall
Pentti Hallapera
R. S. Hallatt
Sam, Alison & Sara Hallatt
Steven & M. G. Hammer
Rudolf & Sonia Hanie
Andrew John Hankey
Miss Susan J. Hansard
Dave & Pat Hardy
Derek Harraway
Errol Harrison
Erich Haugg, Muehldorf
Tielman & Ester Haumann
Peggy Heard
Tessa Hedge
Geoffrey A. Hedgecock
Helen Fowler Library, Denver Botanic Gardens
Dr D. G. Herbert
Basil Hersov
Sybil G. Hewitt
John & Ruth Higgins
Albert Hofman
Max Holmes

Alan J. Hunter
Ian B. Huntley
Erika Huntly
Friedrich Hüttel, Vienna, Austria
Ulrich Hüttinger

Professor Dr H.-D. Ihlenfeldt
Institute for Plant Conservation

J. S. Gericke Library
Harold & Kiki Jackson
John H. Jackson
Mary Jackson
Dr N. H. G. Jacobsen
Elodie Janovsky
Jardin Exotique de Monaco
John G. Joannou
Peter Johnston
Dr H. A. Jonkers
Alarik Jooste
J. Jordaan
Paul A. M. Joubert
Anthony Jukes
Stephen L. Jury

Hillel M. Kahn
Sidney H. Kahn
Stan & Desiré Kantor
M. F. Keeley
W. C. Keen
Mr H. C. Kennedy
Bob Kent
Claud Keytel
Michael & Rosemary Kimberley
Charley Kúnzang King
Michael W. King
Mr & Mrs B. J. Kirby
David J. Kirkbright
Edmund Kirschnek
Anna Kirsten
Ria Kirsten
Gérard Koudouradjian
Daryl Koutnik & Rana Tayyar
Professor Rolf-D. Kratz
Chris & Val Kühn
Michelle Kuun
Edna-Mae Kyriazis

D. J. la Cock
Eric & Isobyl La Croix
Michel Lacoste
Joan Laubscher
Nico en Maud Laubscher, Stellenbosch
John Lavranos
Ernest J. Lazenby
Reinhard & Emma le Roux
Wicus Leeuwner
Helen Leibel
John Lennard
Jean-Marie & Astri Leroy
Letaba Arts & Crafts
Lifestyle Garden Centre
Denzil A. J. Little, MBE
George Lombard
Gaynor Louw

Josua Louw
Professor Roy A. Lubke
Kenneth Hotspur Lund

Kathy Macwhirter
Bryan Makin
C. W. Malan
Clifton Marais
Michael Marais
Gerhard en Rykie Maritz
Ralph R. Martin
Gerhard Marx
Daniel Masson
Victoria Matthews
Phillip & Susan Maxwell
Harry Mays
W. P. J. McClure & Mrs Una McClure
Gavin McDonald
Candace E. McIntosh
Derick McKenzie
M. R. McNeil, Ofcolaco
Kent Meline
Pierre, Irene & Julie-Mari Meyer
Ruth & Walter Middelmann
W. R. G. Millar
D. M. Miller
M. Barry Miller
V. S. Miller
Dorothy M. Minors
J. L. Morris
David Morrison
John Mortimer
K. V. Mortimer
Mrs E. E. Moulder
René & Elmarié Mulder
George Müller
Gordon L. Munday
R. M. Murray

Valerie Neethling
Ivor Guillaume Nel
J. P. Nel, Kimberley
Louis Owen Nel
Karl-Heinz Neuwirth
New Mexico Cactus & Succulent Society
Professor L. E. Newton
Geoff, Lynne & Douglas Nichols
I. T. Nickless
Angela Niebel-Lohmann
Franklin J. Niedz
D. L. Nielsen
Phillip & Helena Nieuwoudt
Hiroshi Ninomiya
Peter W. Noble, Barnburgh, England

D. J. Offord
I. D. Oliver
M. J. Oliver
Willie & Sandra Olivier
Cora & Cyril Ovens
Lindsay & Eileen Owen

Werner A. Paetzold
Dr. med Rainer Pampel
H. Peel

C. W. Pert
Eric Pounder
Ken Price
V. L. Pringle

Lorraine & Roger Raab
Arthur Rabson
Asok Rajh
R. F. Raymond
Mrs F. Reay
Norbert Rebmann
Mr Anthony Roy Reedman
J. A. Retief
Philippe Richaud
Sean C. Roberts
Cedric Roché
Dr Werner Roehre
Dave & Uelah Rosenberg
J. P. Rourke

Maurizio Sajeva
San Jose Cactus & Succulent Club
Rachel & Rod Saunders
Ute Schmiedel
Colin Schroeter
Rudolf Schulz
Paul F. Schulze
Michael H. P. Schurr
Martin R. Schwellnus
Joseph Robert Scott
B. A. Scowen
Professor & Mrs S. C. Seagrief
P. L. Seed
Norman C. Segal
J. R. Selvan
Dr Aleck Selzer
Robert William Semple
Hildred Senger
Docteur R. Sévin, Lausanne
André Share
Pierre & Hazel Singery
Mr & Mrs D. W. Skawran
S. G. Slack
Ansie & Dennis Slotow
C. M. Smith

E. M. Smith
Johan Smith
Keith & Dorothy Smith
Rowena Smuts
Deirdré Anne Snijman
A. N. Sperryn
Heinz Staude
F. E. Steffens
Rob & Sylvia Stein
Dr Elzabé Stoffberg
Sukkulentensammlung, Zürich
Quentin Sussman

Allan Tait
Andrew Tarr
Norman Tate
Colin & Ann Tedder
Bud Terrell
The Garden Shop, Kirstenbosch
Bryan & Jan Thomas
V. G. Thomas
Keith Thomson
Jennetta Tilney
Top Turf Sun City '93
Derek Tribble
Dave & Jo Trickett
J. W. D. Turner
Mrs & Dr J. Tyrrell
Dr J. C. Tyrrell

University of Stellenbosch
Julie Uttridge

Dick & Liz van der Jagt
M. van der Mark
Troos van der Merwe
Paul & Gerda van der Reyden
K. P. van der Riet
Johan & Annatjie van der Sandt
Anne & Errol van Greunen
J. R. P. van Hoey Smith
Albert van Jaarsveld
Professor F. A. van Jaarsveld
J. E. van Jaarsveld
H. van Kerken

Richard Davison van Niekerk
M. & M. A. van Rijswijck
Elizabeth J. van Rooyen
Brian & Jane van Wilgen
Braam & Elsa van Wyk
Rhoda van Zijl
Mev. A. M. S. Venter
Fanie & Julye-Ann Venter
Dr H. J. Venter
Kirsten, Dirk & Carl Venter
Kobus Venter
M. M. Venter
Donvé Vlok
Dr L. Vogelpoel
Dieter von Willert
Dr & Mrs P. Vorster
Lena Vroom

Graham Walker
John Walker
Mike Warren
Hitoshi Watanabe
P. Watson
Professor Ian B. Watt
Ms W. G. Welman
Roland A. Welsbie
John Welton
Paul Whitehouse
Brian Wicken
Stephen Wilhelms
Ann Williams
Huw & Carol Williams
Sam & Dorothy Williams
Jean Wilson
Hilda & Jürgen Witt
Rob & Janet Wood
Mr Eddie Michael Wright
A. K. Wydera

Dennis Young

Michael Ziegler
Ben J. M. Zonneveld
Mr & Mrs Stanley Zway

FERNWOOD
PRESS
P O Box 15344
Vlaeberg 8018
South Africa

Registration no. 90/04463/07

First published 1994
Reprinted 1995

Design by Wim Reinders and Associates, Cape Town
Dustjacket and title page design by Abdul Amien, Cape Town
Production control by Abdul Latief (Bunny) Gallie
Typesetting by 4-Ways DTP Services, Cape Town
Reproduction by Unifoto (Pty) Ltd, Cape Town
Printed by Tien Wah Press (Pte) Ltd, Singapore
Collectors' edition bound by Peter Carstens, Johannesburg
Standard edition bound by Tien Wah Press (Pte) Ltd, Singapore

ISBN 1-874950-04-0 (Collectors' edition)
ISBN 1-874950-01-6 (Standard edition)